ABOVE AND BEYOND PALESTINE

JERUSALEM : THE DOME OF THE ROCK, ON THE SITE OF SOLOMON'S TEMPLE.

ABOVE AND BEYOND PALESTINE

AN ACCOUNT OF THE WORK
OF THE
EAST INDIES AND EGYPT SEAPLANE
SQUADRON
1916—1918

By
C. E. HUGHES
with illustrations by the author

&

*new biographical
afterword by*
IAN M. BURNS

LITTLE GULLY PUBLISHING
2025

First published in 1930 by Ernest Benn Limited, London.
This second edition faithfully reproduces the original
text and illustrations. An afterword provides
biographical details of the author.

© 2025 Little Gully Publishing. The original text is
in the public domain. The typographical arrangement
and afterword are copyrighted by the publisher.

ISBN 978-1-7636268-6-7 (paperback)
ISBN 978-1-7636268-7-4 (ebook)

littlegully.com

 A catalogue record for this book is available from the National Library of Australia

CONTENTS

	Page
LIST OF ILLUSTRATIONS	vii
MAPS	ix
NOTE	xi
I. BUSINESS IN GREAT WATERS	1
II. SHIPSHAPE	23
III. THE LIE OF THE LAND	47
IV. A ROUND OF VISITS	66
V. THE COAST OF ARABIA	86
VI. ON DUTY AND OFF	104
VII. THE LOSS OF THE BEN-MY-CHREE	136
VIII. SOMEWHERE EAST OF SUEZ	156
IX. BEIRUT AND DAMASCUS	181
X. GAZA AND JERUSALEM	203
XI. ON LEAVE	223
XII. THE METAMORPHOSIS AND THE END	261
BIOGRAPHICAL AFTERWORD	269

LIST OF ILLUSTRATIONS

	Page
JERUSALEM: THE DOME OF THE ROCK *Frontispiece*	
MARSEILLES	2
ZEA, IN THE GREEK ARCHIPELAGO	8
SALONIKA	12
OLYMPUS	16
ALEXANDRIA FROM MUSTAPHA	18
SIDI GABER, ALEXANDRIA	21
STATION CAMP, PORT SAID	23
H.M.S. BEN-MY-CHREE AT PORT SAID	24
SEAPLANE ISLAND, PORT SAID	27
SUEZ CANAL COMPANY'S OFFICES, PORT SAID	30
TRAWLERS, PORT SAID	34
TAURUS MOUNTAINS FROM THE GULF OF ALEXANDRETTA	36
ASKALON	50
TYRE	51
SIDON	53
ACRE	54
H.M.S. ANNE	58
GAZA	62
MOUNT CARMEL AND HAIFA	67
EL FULE	68
HAIFA	74
RUINS AT RAMLEH	79
TUL KERAM	81
H.M.S. ANNE	100
R.N.A.S. ARCHITECTURE—1	106
R.N.A.S. ARCHITECTURE—2	107
BEDOIR AND BOOTOIR	110
PORT SAID, THE SUEZ CANAL	115
COALING, PORT SAID	117
PORT SAID: THE MARKET	119
NILE STREET, PORT SAID	120
ARAB TOWN, PORT SAID	126

LIST OF ILLUSTRATIONS

	Page
NAVY HOUSE, PORT SAID	130
THE HARBOUR, CASTELORIZO	140
H.M.S. BEN-MY-CHREE	150
ARABS HAULING IN A SEAPLANE	162
H.M.S. RAVEN II	166
LEBANON	180
AT BEIRUT	182
DAMASCUS: THE STREET CALLED STRAIGHT	188
BEIRUT HARBOUR, WITH SUNKEN TURKISH GUNBOAT	192
SKETCH MAP OF BEIRUT HARBOUR	195
ENTRANCE TO BEIRUT HARBOUR	197
H.M. LANDSHIP WAR BABY, GAZA	206
GAZA	210
BRIDGE AT DEIR SINEID	213
H.M.S. CITY OF OXFORD	216
TUL KERAM	221
RAS EL-AIN	222
EL-TALIBIYEH AND THE PYRAMIDS	224
THE BLUE MOSQUE, CAIRO	228
FOUNTAIN AND SCHOOL OF ABD ER-RAHMAN, CAIRO	232
IN THE SHARIA EL-GHURI, CAIRO	234
THE HOUSE OF THE JUDGE, CAIRO	237
IN THE KHAN EL-KHALILI, CAIRO	242
GATEWAY IN THE KHAN EL-KHALILI, CAIRO	245
BRASS-WORK, CAIRO	249
OLD CAIRO AND THE CITADEL	254
FOUNTAIN OF SHEIK MOTAHHAR, CAIRO	258
H.M.S. EMPRESS AT PORT SAID	262
MEJDEL YABA	265
PORT SAID	268

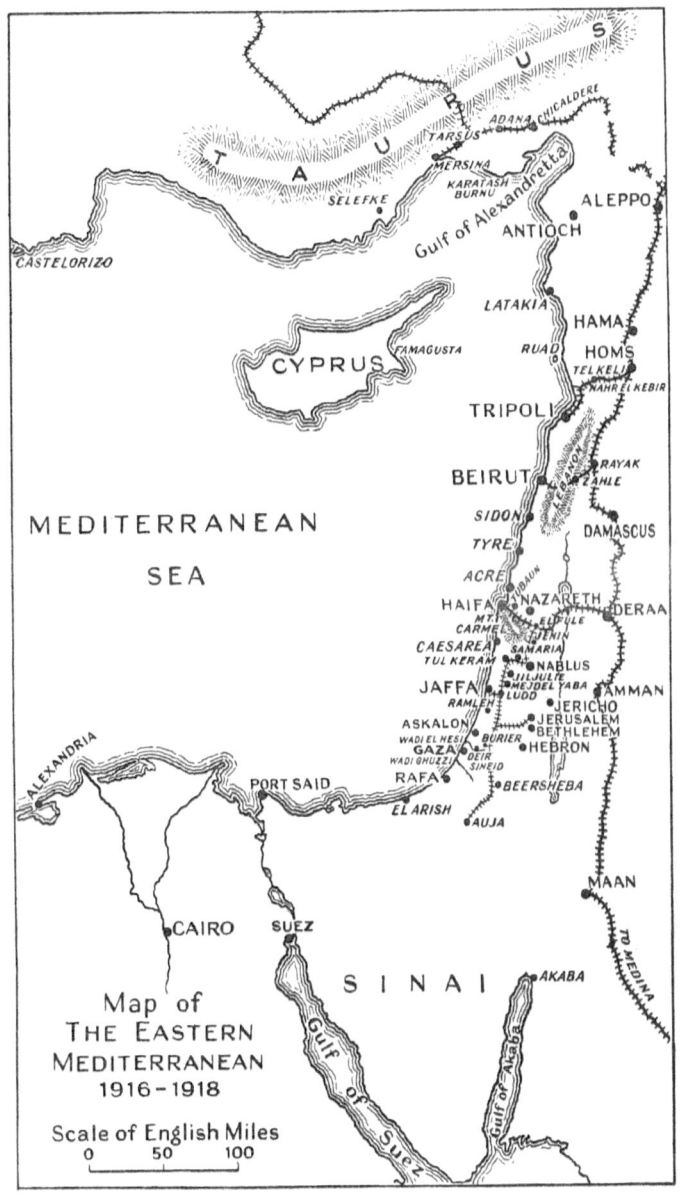

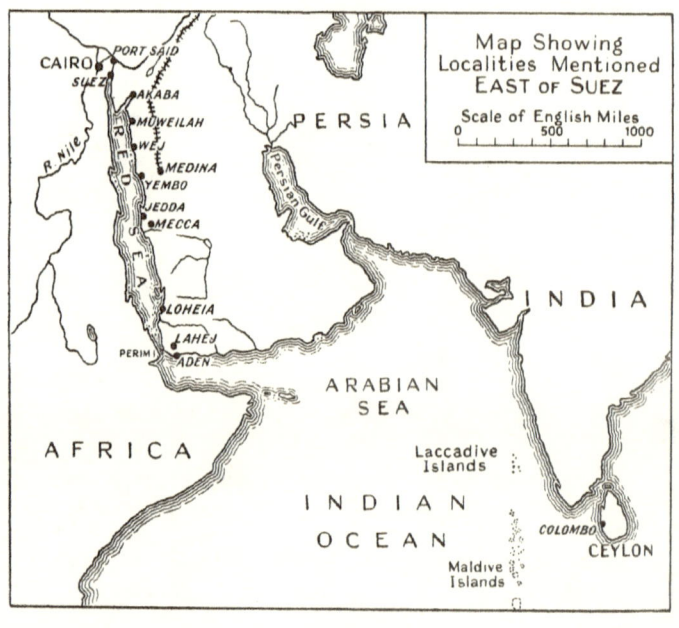

NOTE

THIS book was written during the closing year of the war, and though there are many parts of it which might be differently expressed to-day I have preferred to leave it, with very slight modifications, in its original form. Our war-time mental attitude towards the enemy is, I hope, a thing of the past, but the fact that circumstances compelled us to adopt it has an abiding and, I believe, an increasing interest. That is one excuse for issuing the volume. Another is that though the work of the East Indies and Egypt Seaplane Squadron was very seldom mentioned in published official reports and its record is accordingly very little known, yet without its story the familiar one of the campaign in Palestine and Syria is incomplete.

Of the illustrations I need say no more than that they are taken from sketches made at the time, and as occasion offered. If they appear to be rather unevenly distributed through the book, an explanation may be found in the service conditions which did not always provide an opportunity.

C. E. H.

ORPINGTON, *November*, 1929.

I

BUSINESS IN GREAT WATERS

It should be safe nowadays to assume that everybody knows something about the work of a seaplane squadron, and about the kind of flying done by seaplanes as distinct from land machines.

For the most part the seaplane pilot and observer in the war had a dull time. A very large portion of their job consisted in flights over dreary stretches of water looking for submarines, and the excitement of seeing one, though in the aggregate it came to a good many lucky individuals, was a comparatively rare occurrence, considering the numbers engaged in the hunt.

A more thrilling part was allotted to a very much smaller section of the service, that of carrying out bombing raids along the Belgian coast, and there were also such enterprises as spotting for the big guns of monitors and scouting ahead for units of the fleet. All these things, with which newspaper readers are tolerably well acquainted, and others less interesting, but no less important, came within the scope of the Naval Wings of the Royal Air Force, and much of the work of the East Indies

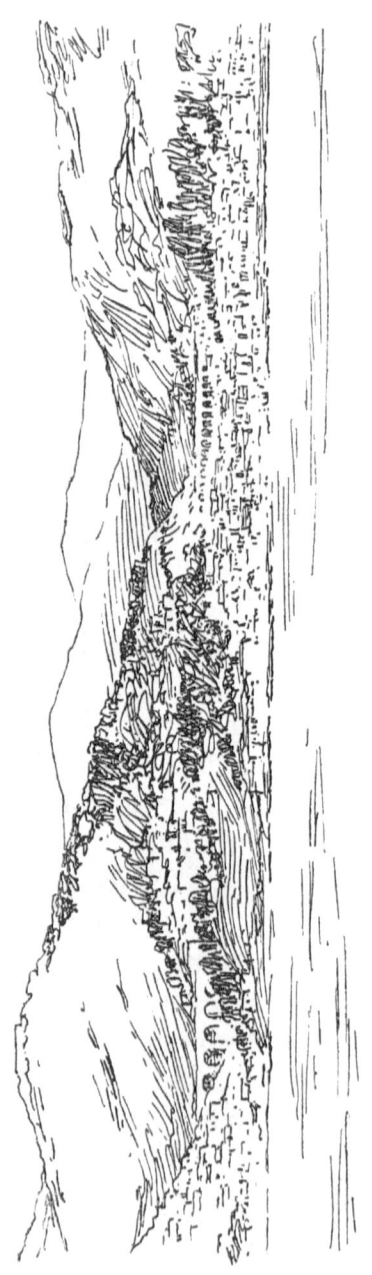

MARSEILLES.

and Egypt Seaplane Squadron (by which name the naval flying unit in the Eastern Mediterranean was known before the R.N.A.S. and the R.F.C. joined forces) could be sorted out into parallel categories qualified chiefly by differences of place and circumstance. But there were ways in which the East Indies and Egypt Seaplane Squadron was unique, and some part of its uniqueness is exemplified by the presence with it from its earliest days of attached military observers.

Regarded from the standpoint of a naval establishment, attached military observers had the appearance of an anomaly, but their presence was justified by the fact that in the earliest periods of its activity the squadron was extensively engaged in overland reconnaissance undertaken for the information of the Egypt Expeditionary Force. From that force, therefore, the majority of the military observers were drawn. They came with varying degrees of experience. Some were gunners with valuable knowledge of artillery emplacements; others were from the trenches; some had special qualifications, as signallers. Few of them had done much flying; none had flown in seaplanes.

Of these attached military observers I happened to be one. I was ordered out direct from England, where I had been since the beginning of the war in training camps and depots doing home service jobs, of which some were not without interest. I had no further acquaintance with seaplanes than such as could at that time have been gathered from the public Press, and that was very little indeed, because the era of reticence had not reached its limit.

A few notes of my journey, typical of that of many others who went out from home, and of my impressions of Egypt and of the station when I joined it, may serve perhaps to indicate some of the differences which existed at the end of 1916 between the actual and the supposed state of affairs.

My mandate came to me one morning in the form of a telegram full of ac-ac-acs from the Director of Personal Services at the War office. I hurried to town, obtained leave pending the receipt of embarkation orders, and overhauled my kit. The orders were delayed for several days, and it was only after the usual series of telegraphed instructions and countermands, that I received a definite notification to proceed through France and report at Marseilles. There was, I remember, a sense of relief that I had not been sent round by the Bay of Biscay.

Nearly a week had elapsed since leaving London in the boat train before I found myself with an accredited share of a three-berth second-class cabin in the transport Ivernia. I had reached the southern French port as one of a party of officers all making for different units, and two of these were my cabin companions. There were no first-class cabins available; a general and his staff and several colonels had had the pick. But such fortunate majors, captains and subalterns as contrived to win prizes in the scramble for what the higher command did not want had merely the questionable satisfaction of close contact with the upper circle. Their bunks looked like iron bedsteads with special sea-legs, but there were four of them in very little more space than that occupied by the three of us in the second class quarters. That means a good deal when your porthole must be hermetically sealed so as to imprison all stray gleams of light.

Comfortably settled then, and provided with a lifebelt, which by the rules of the ship never left our hands, we prepared to face the vague uncertainties of the submarine peril in the Mediterranean, and listened cheerfully one November evening in 1916 to the noises which indicated that the vessel was getting under weigh. I was up early next morning to see if by chance there were any distant view of Corsica or Sardinia. The view was not distant at all, and it was strangely familiar. There, close at hand, were still the harbour and hills of Marseilles.

The ship, having made a promising start, had put back again in the night, and where she anchored she stayed two days.

It was the first direct reminder of "Fritz" under water. The English Channel had been nothing. The escort of destroyers, which certainly indicated possible submarines, seemed like an impassable barrier. They recalled the Israelites passing dry-shod between the pent-up waters of the Red Sea. But at Marseilles there was a taste of the real thing, and a taste rendered not more palatable by the versatility of the deck-talk explanations of the delay.

These explanations had a wide range. There was that of the rather weather-beaten elderly infantry subaltern who, after an incoherent history of his own army record, proving that he ought to have been a major, whispered the confidential information that Germany had got wind of this particular trip, and all the submarines in the Mediterranean were lying in wait a few miles out at sea. That marked one extreme of conjecture, though it was expressed less as conjecture than as fact which could be substantiated, were it not that reliable and authoritative informers might be compromised.

At the other extreme was the far more acceptable theory that the sea was too rough for the destroyer escort. The sea didn't look very rough, and if a sea of that sort were too rough for destroyers, then, surely, the ship must wait for a smooth passage. Which was all to the good. No one ever heard the real reason for the delay—no one, that is, below the rank of field officer, and possibly even they were not all informed. It was enough for the "other ranks" to know that the matter was in good hands, and, sure enough, on the third morning after the original start, we woke to see the sun lighting the snow-capped peaks of Corsica.

BUSINESS IN GREAT WATERS

The period of delay had been spent in alarm parades, at which after a few rehearsals everybody took station quickly and correctly near his allotted boat. The chief business of these parades was instruction in the proper wearing of lifebelts, and it became evident that but for this instruction hardly any one but the ship's officers and crew would have floated the right end upwards if they had been subjected to sudden immersion. A cork lifebelt looks an easy thing to put on, and it is; but you have to know the way.

Practice parades, more and more efficiently carried out as the voyage progressed, were the one nightmare of existence in that very comfortable ship, in which excellent concerts, reading, cards, and pleasant deck lounging, very much as though there were no war, relieved the tedium. There were, it is true, a couple of battalions of troops on board—they supplied the talent for the concerts, and their officers had their duties—but we detached officers, going to special jobs, were very little concerned with anything but our own amusement, though we took a turn with watches.

On the second day from the actual start, the ship lay for a few hours—long enough to get a glimpse of seaplanes there—in the harbour at the southeast of Malta. There were hopes of getting ashore, but nothing came of them. The officers bound for Egypt had an additional hope that they might be able to tranship and get a more direct route, since it seemed almost certain that the two battalions on board were bound for Salonika. These hopes, too, came to nothing, and they were all part of the curiously vague uncertainty which filled the air. Those, doubtless, who sat in high places knew the ship's destination, but others could only guess. To follow the course on the chart hanging near the head of the grand staircase was but to add to the

ZEA, IN THE GREEK ARCHIPELAGO.

bewilderment. The white wake at the stern marked innumerable zig-zags, and they were trivial in comparison with the giant zig-zags which seemed to leave no piece of Mediterranean coast outside the possible range of one's vision. But such stretches of coast line as really did come within range gradually became recognisable as islands of the Greek Archipelago, and one of the concert artistes threw off a hasty parody of " The Isles of Greece " which when he recited it in the evening was accepted as the original.

BUSINESS IN GREAT WATERS

It was on the sixth day at sea, very early in the morning, that Salonika came into sight, and the whole of that day was one of relief that the doubtful waters sprinkled with islands to the south of the little old town, waters in which two days had been spent, were safely traversed.

For it was at a time, one must remember, when to most people in England the attitude of Greece was a profound mystery. One knew of a big Allied Expeditionary Force—that vague, comprehensive term—landed at Salonika, but how much of Greece was friendly to it and how much hostile one did not know. One knew that its lines of communication with sources of supply stretched away out to sea, and looking at the map one saw the intricate clusters of land spaces, each, for aught one knew, a submarine base formed in anticipation many years before the war. To the person at home a voyage through these waters seemed a thing of infinite peril, and the sense of danger felt by those making the journey for the first time lost nothing in intensity from the well-trained alertness maintained in the ship. One might reasonably feel relieved to have it done with, and the relief helped considerably to alleviate the irritating delays and embarrassments attending the disembarkation of those two battalions of troops.

That, at least, is how the detached military officers looked at the matter. The first stage of the sea journey was over, and for variety there was the prospect of numerous interviews with military landing and transport officers, much-harassed folk to judge from those at Marseilles, and the immediate question of accommodation ashore till a ship could be found to complete the passage to Egypt.

But all these difficulties in prospect were overshadowed at the moment by the seeming impossibility of getting ashore at all. It was not that there was any lack of orders. There were far too many of them. Elaborate details respecting a steamer which would land the unattached officers were no less plentiful than statements as to the times at which it was due to come. When it did come it was not merely unexpected but palpably different from any of the descriptions of it. There was, indeed, one essential particular in which it differed. It waited three hours alongside packed with troops and officers, and made not the slightest effort to fulfil its duty of putting them on dry land; and when finally it began slowly to carry out this work, it omitted to take the heavy baggage which had been hoisted from the hold in readiness for it, after many struggles and much labour of identification.

Ashore on the quay was an almost desperate M.L.O. who emphasised the fact that it had been raining by standing in a puddle which reflected his height downwards towards a dark infinity. The more visible and tangible part of him carried a notebook, and faced the insoluble problem of conveying umpteen unattached officers to a distant Rest Camp in motor transport of which the capacity was less than one-third of umpteen. I recalled the fact that I was a military observer, and rapidly exercised my maturing powers of observation. Depositing my hand baggage in a dark corner, I clutched in the crowd a familiar arm—it was that of one of my two cabin companions—and made off townwards. We agreed that we could take our turn later with that motor lorry.

Luckily, I had chosen better than I knew. My companion had been to Salonika before, and had some acquaintance with its desperately ill-paved streets, and its uninteresting-looking but not badly equipped restaurants. We were able to get dinner, and after it we returned at leisure to the quay. The group of unattached officers had perceptibly thinned, and there was no M.L.O. Seemingly he had left the problem to solve itself. Only temporarily, however. Before long, he was back again, and a panting motor lorry, evidently hoping that the next journey would be the last, stood by for his orders. Words failed him. He indicated the big but too hopelessly small vehicle in the gloom with a gesture which said unmistakably, " That's all I can do for you."

Seats were taken, and there was yet unprovided for an incompressible group. Then some spokesman put forward a proposition. " What about dossing on the steamer?" The M.L.O. leaped at it, and

SALONIKA.

there was a quick grasping of the opportunity. But two officers still remained. I was one, and I boldly suggested an hotel. "Certainly," said the M.L.O., "if you can find one. They are all packed to the teeth."

Once more, this time with a new companion, I ventured forth into the town, and luck favoured us again. Not at the first attempt, nor at the second, but finally, a room was found, engaged, and occupied. And the pair of us finished the evening at the Pictures.

Next morning, awakened by the cries of the Salonika bootblacks, who ply literally a roaring trade on the kerb of every street that has a kerb and in the roadway of those which have not, obviating entirely the need for any domestic service in this particular, we sniffed the atmosphere of a strange and cosmopolitan city. It combined Roman antiquities, amazing squalor, mediæval town planning, and twentieth century picture palaces. It was also the base of Allied operations, of which the nearest point was some seventy miles away. The ingredients were mingled in an indescribable patchwork of sights and sounds, and among them all, as it were interwoven like some geometrical pattern in a variegated fabric, the military machine pursued its slow, certain, and relentless industries.

It was not until our eyes had begun to discern the rhythm of the seeming confusion that we realised something of the task that had confronted the harassed M.L.O. of overnight, and admired the coolness with which he carried it out. Troops in battalions and squadrons were being hurried forward, and officers unattached at Salonika, however firmly they might be attached when they reached their real destination, were to the M.L.O. in the nature of overtime.

It was rumoured that even our fellow passengers, the general and his staff, had passed the night in an unfurnished empty house. The personal baggage of unattached officers was a secondary matter compared with that of complete units timed to arrive fully equipped somewhere inland at a specified moment. The general's baggage, despite the frantic efforts of a brilliantly-tabbed youthful A.D.C., had remained in the hold of the Ivernia, and almost our consciences smote us that we had slept in beds.

After breakfast in a restaurant peopled with numberless unidentifiable uniforms, for all Allies seemed to have a footing in the town, we splashed through the mud to the office of the M.L.O. We had previously, out of our new-born respect for the M.L.O., purchased the services of a boot-black, but it was money thrown in the gutter. In Salonika, after a shower, no one can walk clean shod for 20 yards unless he be an expert on stilts.

The M.L.O. believed—he was not sure, mind ; how could any one be sure of anything relating to unattached, attached officers?—he believed that the heavy baggage would be brought ashore during the morning. Better report again at 2.30. Meanwhile, there was no chance that day of a ship to Egypt. Anyhow, report first thing to-morrow.

So the morning and afternoon were spent in a quiet contemplation of strange sights, sounds and smells. Periodically excursions were made in search of the missing baggage, but always without success. We learned in the course of these wanderings that old Salonika, partly walled and dotted with many minarets, and clinging carelessly to the side of its hill, spread east and west into something more modern and orderly. One way there were dockyards, the other pleasant-looking residences. We made acquaintance, from the heights, with the superb distant mass of Olympus and the winding coastlines, and found in the grimy heart of the town itself—much of it since destroyed by fire—that tram lines could be laid and trams run through streets which had to be crossed afoot, as one crosses an English brook on stepping stones, by springing perilously over the gulfs of mud between huge cobbles, relics possibly of a roadway which had had its prime in the dawn of the Christian era. It was a weird chaos of the immemorial, the historical and the actual. In that leisurely sauntering day, with the mechanism of war all around us, we almost forgot that there was a war.

But the next day brought it home to me, if not to my companion, who had left me on some quest of

OLYMPUS.

his own. The M.L.O. had, it seemed, managed to snatch an unoccupied second. There was a pause perhaps in the progress of troops to the front, a longer break than usual, possibly, between two battalions on the march. However it might have been, he, or his near companion, the Military Transport Officer, had found a ship bound for Alexandria. Accommodation in it was limited, and it was a case of first come, first served. My name was put down for it.

Then, suddenly, I remembered my missing baggage and sped forth on a round of inquiries. They revealed the fact that it had been brought off from the ship, but they did not reveal what had become of it. It was nowhere visible, and one was reduced to conjecture. It must have been taken to the Rest Camp, whence no human effort could extricate it in time to catch the outgoing steamer. It was the moment for quick decisions. Full of respect for the Military Base Staff, a respect born yesterday and growing to-day in intensity, I took the risk of leaving my heavy kit behind. Giving a full description of it to the Military Forwarding Officer, I drove to my room, settled the bill, collected my hand baggage, and caught the ship. It should be said that I did this in the face of the advice of the M.L.O., the M.T.O., and the M.F.O., but I believed in them, and events proved me right. Three weeks later the missing bundles followed me to Egypt.

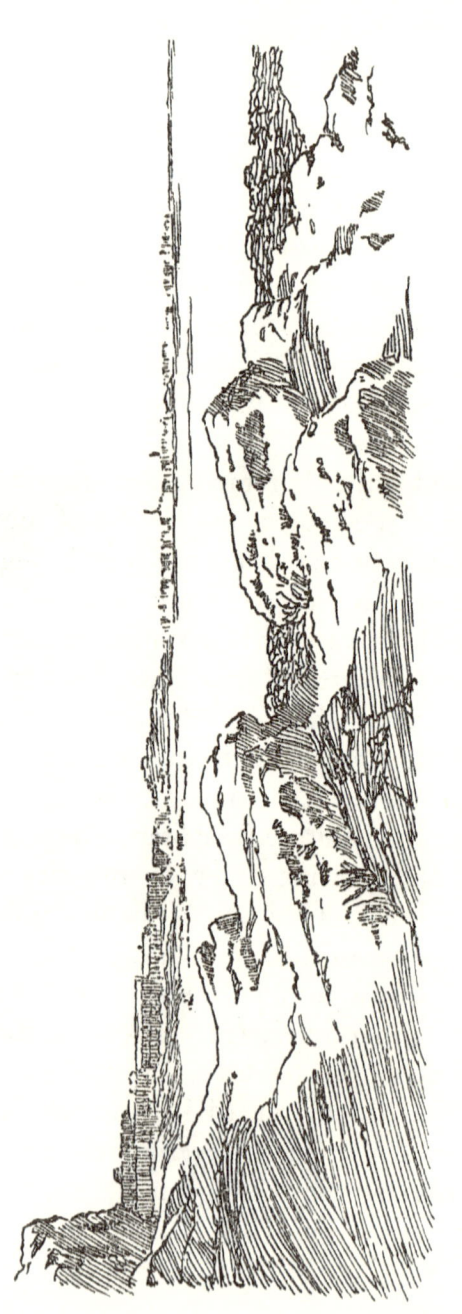

ALEXANDRIA FROM MUSTAPHA.

BUSINESS IN GREAT WATERS

The new ship was a small comfortable packet whose passengers were a select few—the more alert—of the crowd of unattached officers who had come out from Marseilles, and half a dozen German prisoners in great terror of torpedoes. Their fears were not without some foundation, only that day " U " boats had been busy near by in the course which seemed to be the direct one from Salonika to Alexandria, and there was further evidence of activity before the voyage was over, though the immediate thrill was experienced only by the night watch. The passengers asleep in their cabins knew nothing until the next day when the hushed alarm provided a welcome topic of conversation.

On the third morning numbers of flying fish, looking themselves like miniature seaplanes as they skimmed the waves round the ship, gave a pleasant hint that the journey's end was approaching. About mid-day came the first glimpse of Egypt heralded by the distant white sails of the fishing boats working in pairs off the coast. Then there appeared the lighthouse above the horizon, and the tops of houses and the low long stretch of land. Finally came the interest and excitement of being piloted through the swept channel. By five in the afternoon the journey was over, and the unattached officers were once more in the hands of an M.L.O.

Receiving instructions to report at a Base Details

camp at Mustapha that evening, and having time to spare, several of us hurried to the telegraph office to cable home news of our safe arrival. We had traversed safely, and by an exceedingly indirect route, the waters of the Mediterranean, and the more intimate knowledge which we had acquired of their perils recalled to mind the feelings in regard to them with which we had left England, feelings which were shared in a greater degree by those whom we had left. We had come through safely, but the dangers, we realised as the events of the voyage were recalled, were none the less there. Escorts of destroyers, harbour booms, friendly minefields, infinite precautions served but to emphasise the fact that few stretches of water in the Mediterranean could be called safe, few be declared free from potential hostility. In England one knew little of the war in these parts. What one heard related chiefly to operations on land, and of these there was little beyond the broad generalisations of official despatches. We had begun to see in detail something—though as yet only a very small part—of the war at sea, and it was good to get one's feet on dry land again.

The Base Details Camp gave a further new conception of the magnitude of the war. It was not merely that the camp itself covered many acres of ground, that it included many vast buildings belonging to some deserted Oriental palace, or that, in spite of this, I was accommodated in a tent. The

BUSINESS IN GREAT WATERS

magnitude of the war forced itself on one's notice, not through any of these things, but through the discovery that no one in the camp had any idea who I was or why I had come there. The Army had ramifications from this base camp which fanned out (in echelon) far away into the desert, east, west and south, but the camp Orderly Room knew nothing of the R.N.A.S. beyond a vague realisa-

SIDI GABER, ALEXANDRIA.

tion of the meaning of the initials, and a still more vague one of the possible presence of one of its units in Egyptian waters.

Letters of authority were produced, examined, and copied; notes were made; signals were sent; and in the meantime, feeling something like Kipling's giddy haramfrotite, soldier and sailor too, I took my turn as orderly officer and censor, and between whiles marched very mixed parties of resting soldiers down to the miniature range and back. Also, between whiles, I had a look at Alexandria, which, but for its few mosques, and the quaint though straight-cut streets of the native quarter near the docks, was too much like a French provincial town to appeal much to a sense of novelty. As appeared later, one must go to Cairo to see something of the real Egypt.

II

SHIPSHAPE

CHRISTMAS was only a few days ahead when the reply came to the signals which my arrival had set going. I received orders to report on board H.M.S. Ben-my-Chree at Port Said, and with a feeling that the bonds of attachment were at length tightening

STATION CAMP, PORT SAID.

towards contact, I took the early morning train which skirted a northward angle of the Nile Delta. Three or four hours with a prospect of rich cultivation, and as many more with little to see but stretches of bare desert and the banks of the Suez Canal, and I was in Port Said driving in a cab over

H.M.S. BEN-MY-CHREE AT PORT SAID.

the fenced-in and closely guarded cobbled precincts of the harbour.

Arrived at the point which was nearest of approach to the ship, I was faced with a difficulty. There she lay, flying the white ensign, 50 yards—I knew nothing of cables' lengths—away. How was I to get there? I was wholly ignorant of the ways of the Silent Service beyond such of its doings as had, by their sheer brilliance, compelled some brief explanatory, half-apologetic paragraph for public consumption. There was no hint there to guide a soldier in his first steps on a quarterdeck, though there was much which might render him shy of taking them.

Did one hail with "Ship ahoy!" or some such phrase reminiscent of Tom Bowling and Peter Simple, and ask to be taken aboard? Or did one hire a rowing boat and boldly invade by means of the accommodation ladder the sacred boards beneath that defiant white ensign? I began by doing the wrong thing. I accosted an R.N.R. "two striper" who was engaged near at hand, and asked what to do. The "two striper" concealed a twinkling eye with hastily formed wrinkles of earnest solicitude: "Sorry my barge isn't here, or I'd put you aboard."

It was possible here to smell a rat. Admirals, captains, perhaps commanders, had barges (according to the book) and this officer hardly fitted the description. I thanked him, took my courage and my baggage in both hands, hailed a felucca, and arrived at my destination.

But not yet was I to feel completely at my journey's end. I was there to report, but the process of reporting demands the presence of at least two individuals, and the second of these was absent. The Commander, I learned, had gone ashore, riding.

There were working hours, then, and nonworking hours for a ship in port. Evidently the ceremony of reporting must be deferred until next morning. Not being, therefore, yet officially attached, and knowing as yet no orders but my own, I set out on an expedition of investigation. I inquired for another military observer, one W., whom I had known some time in the almost forgotten days of peace, and was directed to the seaplane Base—"That island over there."

Gazing along the extended arm and forefinger of the informant, one could see in the gathering twilight distant land on the far side of the Canal, and a group of huts, but nothing very greatly resembling an island, if the old definition of a piece of land surrounded by water held good. The pointing finger, becoming more explicit, indicated a flagstaff, a very unnaval looking flagstaff. I was told to make for that. One of the ship's boats carried me. The island looked more like an island at close quarters, and I learned that W. had also gone ashore. Possibly he might be found at the Eastern Exchange Hotel.

SHIPSHAPE

I made for the nearest point of the mainland, and set out to find the Eastern Exchange Hotel. If you look at a street-map of Port Said, you will say that there could scarcely be an easier town to find your way in. The streets are all straight and the corners mostly square. But place yourself, a stranger, in the dark, without a map, at a distant spot on one of the quays which stretch far along the banks of the Canal in the direction of Suez, and from there try to find your way. Walking, mind; there were no

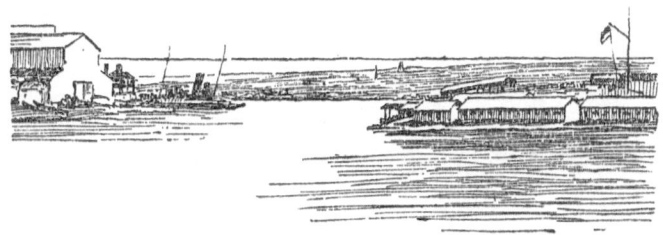

SEAPLANE ISLAND, PORT SAID.

cabs to be had. I took some credit to myself that I did it in an hour. I arrived at the hotel with a fixed and erroneous idea that Port Said was the most ill-designed and most irregular town on the face of the earth. But I did arrive there, and W. greeted me as one did in those days greet such stray acquaintances from distant England.

After that all was plain sailing. Dinner at the Casino, a walk to the Customs House, a boat up the Canal in the moonlight among the black hulls of moored vessels, with the distant buildings of the town looking strangely like Venice, and we were again at the island. The long huts seen earlier from the ship resolved themselves into habitable dwelling-places. Inside they were partitioned into cabins, not over roomy, but very inviting to one who had been on the move for many hours.

The next morning began a series of rapid impressions of many new things, impressions which the process of time alone sorted out in their true perspective. At the moment they merely bewildered one with a mass of detail, co-ordinated with a central idea, that of a weapon of war, quite strange and wonderfully capable of hard hitting; a weapon in the wielding of which I was thenceforward to bear a part.

For the process of attachment had been completed. The dread ceremony of reporting on board one of His Majesty's ships had shaped itself into an affair of extreme simplicity in the preliminaries, followed in the event by a few minutes of affable conversation with the Squadron Commander, a flying officer whose name had been in everybody's mouth since the earliest months of the war, and who is now an Air Commodore.

It was during this visit that I was able to note a point which had not previously made much impression on my unnautical mind. The Ben-my-Chree looked like a cross-channel steamer, except that behind her funnels was a big square barn-like erection occupying the width of the ship. This was the hangar where the seaplanes were housed, and I learned for the first time of the existence of seaplane carriers. It was revealed that the work of the military observers was conducted in seaplanes which were carried in the Ben-my-Chree or other ships to places at sea far distant from Port Said.

Back on the island there were certain definite duties to take on. The Military Unit attached to the R.N.A.S. had an Adjutant, who was an office rather than an officer. It was quickly apparent that no one could be sufficiently certain of residence on land to feel justified in assuming the duties in permanence. The military observers, therefore, had been, so to speak, standardised. Each one of them had a turn, and was assumed to be capable of identifying himself with the Adjutant whenever circumstances demanded it. Any new hand began forthwith to undergo the process of standardisation. I found, moreover, that my education was largely a matter of adjutancy self-taught, for though I came upon the names of many other observers in the mass of papers through which I waded, there was none there in the flesh excepting W., who, being O.C. Observers and combining those duties with those of Intelligence Officer, had other things to occupy him.

The Intelligence Office, a place of amazing organisation (described by unbelievers, so one learned later, as hot air), had many wall maps

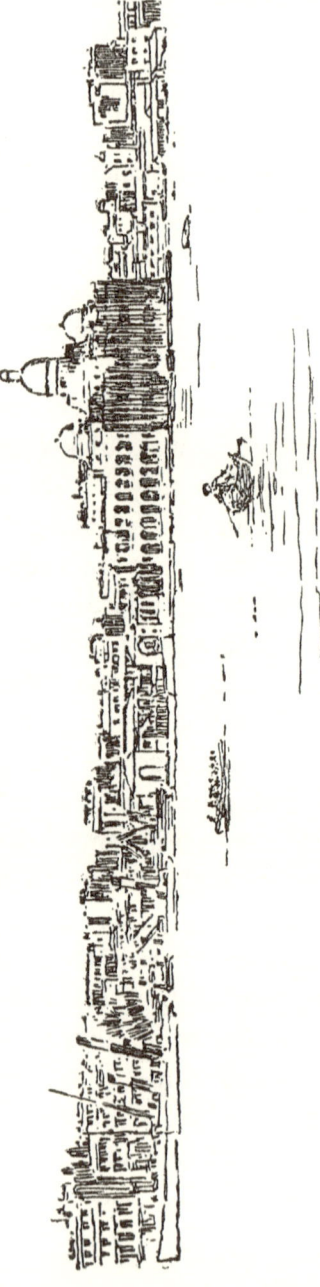

SUEZ CANAL COMPANY'S OFFICES, PORT SAID.

which revealed details of the Turkish positions hitherto undreamed of. These took more than a day or two to digest—every observer was expected to digest them—but it was not these so much as the absence of the other observers that showed that something was afoot. Discreet inquiries of W. gave some light. The Ben-my-Chree was the chief of the three Seaplane Carriers attached to the station. The Ben-my-Chree, in which I had reported, was then lying in sight of the island. One of the others was there too; but the third was—where? There was no answer to this. She was at sea behind an impenetrable curtain of mystery. In due course, the story of her work would be visible and legible, though still secret, in the tangible form of reports and photographs filed in the Intelligence Office. Until then—silence.

One recalled the notices, touched with the inherent artistry of the French nation, which one had read on the journey out. *Méfiez-vous*; *taisez-vouz*; *les ennemis ont oreilles*. There were Egyptian, Arabic and Turkish ears, as well as German. It was the first lesson in standardisation.

The second lesson followed quickly. On the afternoon of Christmas Eve I was told that there was an expedition in prospect and that I had been detailed to join one of the two ships which would be engaged. My work would be to take charge of the intelligence and to superintend the production of a report of the flying operations carried out from this ship, H.M.S. Raven. Stated in the curt terms of an order, the commission seemed to assume that the process of standardisation was already complete, that though I hardly yet knew my way round the station, I was fully capable of carrying through the task which, equally with the duties of adjutant, might fall to the lot of any observer at any moment. There seemed to be here something of an implied compliment, though doubtless it was accidental.

The duties from that moment forward were definite; laid down, in fact, in black and white. The stationery required for such trips had a specially designed box, a wonder of compactness, which was kept stocked in accordance with a list. Ink, pens, pencils, writing paper, drawing paper, sun-printing paper, tracing paper, drawing instruments—everything had its place, and there was no space to spare, vacancies indicated an omission; a false balance of repletion in any department meant that the lid of the box, which formed a drawing board, would not fit into its place. Swollen with importance, and with the assistance of an exceedingly capable N.C.O. draughtsman, I got the box packed and the lid into position and sealed up.

So much for the stationery box. Next the intelligence box, a far more exciting matter, because the filling of it rendered necessary some knowledge of the nature of the forthcoming operations. The scheme as revealed was to make a bombing attack on the railway bridge at Chikaldere, a spot some miles inland from the Gulf of Alexandretta, in the north-eastern corner of the Eastern Mediterranean. The Raven would sail during the morning of Christmas Day; the Ben-my-Chree, whose greater speed allowed her the luxury of a Christmas dinner in port, would start on Boxing Day. They would meet on the following morning and—get to work. There was no avoiding the feeling on learning this that one was getting at length into touch with a piece of the real war absolutely unknown to the public at home.

Meanwhile there was the intelligence box to fill. Files relating to the Alexandretta district, index cards with the essential information boiled down to a minimum, books, maps, plans, photographs, were collected and stowed into their allotted spaces in the box, and that in its turn was closed down. The work was new, and it took longer to do than it takes to tell.

It was done while elsewhere on the station preparations were being pushed forward for the perfection of other parts of the fighting machine. Officers in charge of engines, armaments, stores, were hard at it; and flying officers, pilots and observers were being collected from the Anne, the third seaplane carrier, which meanwhile had unostentatiously arrived at her moorings. She had been for an extended cruise in the Red Sea, and some of her officers had contrived to get leave which should have extended over Christmas. Telegrams to Cairo recalled them, for Chikaldere, seemingly, had developed suddenly and unexpectedly the symptoms for which bombs were the recognised remedy. The trains into Port Said late that Christmas Eve carried several crestfallen faces.

Christmas morning saw everything and everybody on board the ship. By noon she had passed the avenue of craft of all sizes which line the Canal, with its sea wall and the statue of De Lesseps,

TRAWLERS, PORT SAID.

dating for all time as mid-Victorian the work of which he was engineer, and by the early afternoon the prospect on every side was bounded by a horizon of Mediterranean Sea. It offered little of interest when it had finally walled from view some distant trawlers which had apparently been trying to destroy a couple of mines with gunfire. That something had undoubtedly been in the neighbourhood to lay the mines appeared, somehow, in a far less menacing light on decks above which fluttered the white ensign, than a similar presence had appeared a fortnight earlier, when I was zig-zagging in a small passenger steamer from Salonika to Alexandria. Then one was as it were entrenched, but entrenched behind scarcely adequate defences; now the position seemed to be entirely reversed.

The ship herself lent colour to this pleasant thought. The first thing that had struck me as I stepped, saluting, from the top of the accommodation ladder, was the fact that the name-plates above the cabin doors were in the German language. H.M.S. Raven II. was a captured ship. When the war began she was the Rabenfels with a pair of big iron crosses on her straight, thin funnel. The story of her taking is neither here nor there. Taken she was, and her after decks were adapted for the disposal and hoisting in and out of seaplanes. A coat of Service grey transformed her outwardly, and the two iron crosses were removed bodily as decorations to the wooden walls of two huts on the island, adjoining that cemented square of sand which was called the quarter deck, for the island, being under naval command, was for all intents and purposes a ship. The Raven was a not uncomfortable vessel, but, being built for cargo and not for passengers, there was no superabundance of deck space, and what there was seemed to

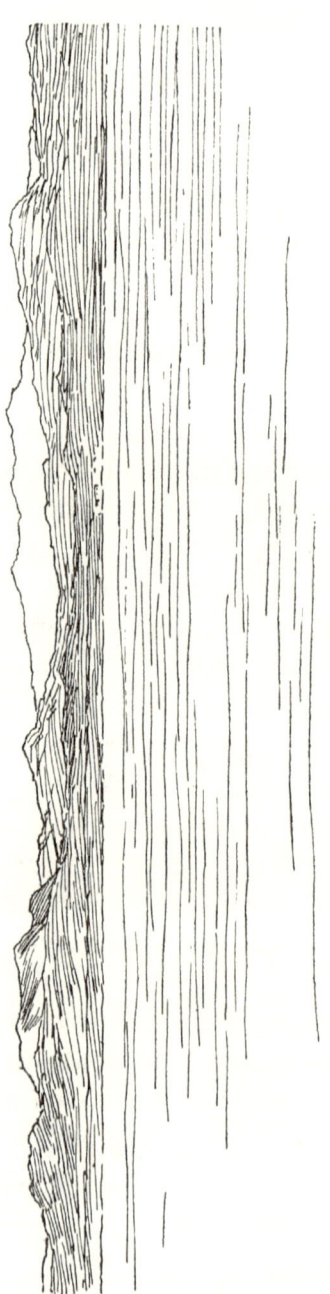

TAURUS MOUNTAINS FROM THE GULF OF ALEXANDRETTA.

be largely encumbered with steam pipes only partially protected. There were two ways of becoming acquainted with this peculiarity when the ship was darkened at night.

As you made your way from your cabin to the ward room, forgetting, if you had already noted them, the positions of these off-shoots from the boilers, you could catch your toe in one and take a toss headlong on the deck. That was one way. Or, proceeding more gingerly and feeling ahead with your foot, you could find your ankle in close contact with metal most unpleasantly heated. That was the second way. During my first evening afloat in the Raven I tried both.

On Boxing Day I was awakened by the sound of guns. In one of those moments of gleaming aberration which sometimes occur when one is only half awake, I suggested to K. E., the Flight-Sub. whose cabin I shared, that the cannonade was evidently in honour of " Bang Holiday." K. E. knew how to deal with such things. He explained quite seriously and at some length, that the ship was loosing off a few rounds for practice. He was not quite so terse, but his meaning was exactly that of the wit who declared that " only dull people are brilliant at breakfast." Duly chastened, I dressed and proceeded with the business of the day, which amounted to very little beyond searching a blank prospect of water, and looking forward to to-morrow.

To-morrow brought early news that the Ben-my-Chree had been sighted, and was rapidly overhauling the Raven. Land had been sighted, too! The ship was in fact entering the Gulf of Alexandretta, and it was not long before it was only from her stern that open sea was visible. On all other quarters, uncomfortably near as it seemed, though really at a safe distance, rose the hilly country of Cilicia, spreading down to a flat coastline, and to the north-west, a distant background radiant with the slanting beams of the morning sun, towered the snowy peaks of Taurus.

The sensation uppermost in the mind was, that the presence of the two ships in those waters was a piece of colossal impudence. The thing was so leisurely, so much a matter of course, so completely tinged with a supreme unconcern. For in the earliest days of the war I, in common with many another armchair strategist, had pictured the landing of a British Expeditionary Force in that very Gulf of Alexandretta. It seemed so obviously the thing to do, to go in there and cut the railway which linked Constantinople with the Mesopotamian and Syrian fronts. Time went by, and this easy route to victory in the war with Turkey was never taken. At home we could conceive of no reason but that the Turks had foreseen the possibility of it, and organised coast defences which made it impracticable. And yet, here were the Ben-my-Chree and the Raven, like a couple of yachts on a cruise, quietly taking their pleasure in waters which were at the very heart of this armchair expedition. I had come from England filled with the idea that all Mediterranean waters had their dangers, but I should have indicated this spot of all others as, if not the centre of the danger zone, at least one in which intruders might look with a maximum of certainty for a warm reception. To go there was surely asking for trouble.

Yet the place appeared peaceful enough. Away on the port bow—I had already got that much of naval terminology, and almost imagined myself back with the battalion giving an order "Company starboard—turn, by the port quick—march"—away to the left, the Ben-my-Chree was already hoisting out seaplanes, one of which could be seen through glasses apparently taking a joy ride over the mouth of the river near by, and the Raven circling in her appointed station, was waiting for orders to follow suit. Flying officers, with their goggles and helmets, maps and haversacks, were standing by (which is much the same as standing to). There was a general fever of alertness, when a look-out directed attention to a strange craft away on the starboard quarter. Glasses abandoned the Ben-my-Chree's seaplanes, and were trained in the new direction. Where previously there had been nothing, there was now a long white line on the surface of the water. Evidently a submarine, though the colour possibly was unusual, for what else could appear so suddenly at a spot well away from the shore and hitherto untenanted ? The Raven's gun crews took action stations. The first shot fell short. The second got the range to an inch but was a little to the right. As it struck the water, the white submarine was seen through the glasses to tumble with a strange fluttering. A moment later it had disintegrated into a flock of startled seagulls.

It was not long after this that the wireless signal came from the Ben-my-Chree to hoist out. Two machines were to go away, and the pilots in anticipation had already run their engines, so that everything was in readiness for the order. I watched from a convenient balcony, so to call it, of the cabin deck, the spry movements of the air-mechanics as each seaplane in turn was slung, lifted from its place on the lower deck, guided overboard by deft hands and long bamboo poles with padded ends, and lowered upon the water. Pilots and observers were in their places, and it was a matter only of a few minutes before they were air-borne, and, gradually rising, shaping course for the distant land.

There was nothing to do then but wait ; nothing in prospect for an hour or more, but that period which is at once for the actual airman of all periods the most exciting in a carrier operation, and, for those left in the ship, the most anxious. In the machine, it is a time of tense alertness for both the pilot and his " passenger." In such operations as these on the Syrian coast, there was always the thrill of exploration. Flights inland, although frequent, were never so much so that one could be sure of one's way. And Chikaldere was not familiar at all. There were hills to cross before the bridge was reached, and it was new country to both the Raven pilots and those of the Ben-my-Chree.

Such information as was known in regard to enemy guns and troops was scanty. Any convenient hill-top might have been lately supplied with "Archies," and, for low fliers, any dark patch on the ground might hold rifles or machine guns. One had to watch out. And there were the engines to think of, too. However well they might have been tried and tested, however well acquainted the pilot might be with the personal idiosyncrasies of his own machine, something, might go wrong. No aerial engine was ever yet absolutely and unquestionably perfect. Moreover one had to watch in, too. For those in the machines there were no blank moments. There was always the stimulus of essential readiness for the probable and the possible. There were always things to do; always things which might have to be done.

But in the ship, one passed that time of waiting in an atmosphere of queries, conjecture, speculation. "Will they do the bridge in?" "Will they get hit?" "Will they come back?" One did not put the last question into words, but it was framed with more or less distinctness at the back of every mind.

One could gauge pretty accurately the length of time during which the machines ought to be away.

As the estimated period drew towards its close, eyes, hitherto indifferent from past experience, gazed more steadily on the line of distant hills, became riveted. And at length the first machine, a tiny speck high in the air, drew into sight. The second was in sight also, before the first had " landed," as, for fault of a better word, one described the process of coming down upon the water, and ten minutes more of impatient waiting found both pilots and observers telling their experiences.

Only one direct hit on the bridge had been scored, and that did not appear to have done very much damage. They had left the Ben-my-Chree's machines still at work. There had been rifle fire, and one observer reported a few bursts of shrapnel, but officers and machines had come back untouched. The pilots had words of praise for their engines as though they were almost human, or at least alive, like a horse that has run a good race, or a favourite dog. The one salient fact, however, that issued from this hubbub of animated conversation, was that Chikaldere Bridge stood where it stood before. It was not destroyed; and the orders had said that flights were to be continued until that end had been achieved.

The day's work was not over. Already the air mechanics were giving the machines a rapid overhaul; the petrol tanks were being replenished; two more pilots and observers were standing by pending the receipt of orders from the Ben-my-Chree in reply to the brief summary of the morning's doings which had been flashed to her immediately on the completion of the two flights.

From the relief observers there were inquiries as to the nature of the target. What sort of a structure was the bridge? How was the wind? What about getting down to three or four hundred feet? The bridge, it appeared, was an excellent target in length, but not in breadth. There were three long spans crossing the river supported on two built piers. You could get the line and have a very fair chance of hitting it somewhere, only it was so infernally narrow. A single line of rails ran over it. Two bombs from K. E.'s machine had fallen into the water just alongside. It was a structure of open iron-work, and even with a perfect aim, bombs might conceivably drop through it. But it was grand fun going for it.

What appears here as a connected, if somewhat staccato narrative, had at the time the form merely of a turmoil of chatter, from which a silent listener was able to gather details as one speaker or another commanded attention with the recollection of some fresh incident or piece of information. In the midst of it arrived the expected signal from the Ben-my-Chree. One seaplane was to be hoisted out to continue the operation. She was on the water with her pilot and observer in their places, when there came a second signal to hoist in and return to Port Said.

The "Ben's" machines, then, had done the trick. Later, when both ships had packed up and were ready for the home run, a further signal reached the Raven. It was couched in what one may call semi-official language, the language which is sometimes used by commanding officers in moments of supreme satisfaction when the veil of their dignity is momentarily lifted, and they stand revealed (to those who look quickly enough) as human beings made of very much the same stuff as their subordinates. The signal stated that four direct hits had been scored. This was the official part. The human part said that the Commander was satisfied that, the squadron could still do a job of work.

After that things went pretty cheerily on the way home. The Ben-my-Chree soon disappeared hull down, and the Raven plodded along with her twelve knots or so, a good second. But my job was not yet done. There was the report to get out, and though it was not a lengthy one compared with many that had been produced in similar circumstances, it was the lengthiest that I had yet tackled. Not that I really had very much of the work to do. The other observers, old hands at it, saw the thing through. Together they worried each other for written accounts of each flight, and harassed the writer who had come with them for typewritten versions of them. Together with plates exposed during their flights they invaded the gloomy spot below decks where dwelt Wilkie the photographic rating, and miracle-worker. Wilkie, formerly a sergeant, latterly a C.P.O., thought nothing (or, at least, said nothing) of "carrying on" when the ship was down the Red Sea and the temperature was about 120 degrees and cold water a thing unknown. If put to it, he could develop a negative in his shaving pot and wash prints in an egg-cup, and, whatever the handicap, he could turn out his part of the report so that it was a joy and a delight to G.H.Q. when they received it.

Together they read, revised, corrected and collated and finally presented to the Flight-Commander for signature, a bound booklet of which I, who after all was nominally responsible, felt inordinately proud. I am not sure that months later, when I had done the same thing many times and with far more of a direct hand, I did not occasionally steal quietly into the Intelligence Office for a peep at this first effort, duly added to the Raven file in the drawer marked "Reports of Flight Units."

Port Said was pushing the tip of its lighthouse above the horizon when I left the Flight-Commander's cabin with the documents signed, and two or three hours brought the ship to her berth. My first stunt was finished, and the island routine was resumed. There followed several pleasant and busy weeks during which we all underwent the process of further standardisation. We learned about bombs and machine guns, and how much easier it was to miss with them than to hit. We took photographs, we studied Morse; we learned codes; we pored over Admiralty instructions. All of which things were very interesting—when you were doing them.

Meanwhile, let us look into the exact meaning of this Chikaldere business.

III

THE LIE OF THE LAND

Some months after the attack on Chikaldere Bridge rumours trickled through giving some idea of what had happened. The bridge had certainly been hit, though not destroyed, but the damage to the single line of railway had been considerable. The damage, moreover, had been inflicted exactly at the time—was it just a strange coincidence?— when the line was to have been used to convey big guns to the Turks defending Bagdad, and the rumour said that the guns never got there.

It is not for me, a mere unofficial gleaner of gossip, to decide whether or not there was any truth in this. Bagdad was taken on 11 March, 1917, and I believe it was doomed anyhow, though I should like also to believe that the R.N.A.S. at Port Said had a share in its fall, however indirect. But the truth of the rumour is immaterial to my present purpose. True or not, it throws a good deal of light on the value of such operations as those in which the East Indies and Egypt Seaplane Squadron was engaged during 1916 and 1917. These operations were always liable to upset the enemy's calculations, very often, they actually did upset them, but whether they did or not they succeeded in creating in the mind of the Turks a feeling of insecurity which made it necessary to keep troops and guns many hundreds of miles behind the front line.

The name of the Squadron implies a wide area of activity, but the actual area over which it worked was even wider. Its seaplane-carriers were frequently in the Red Sea, and one of them cruised for some months in the Indian Ocean, but it is not in this direction that the boundaries indicated in its title must be stretched. Egypt in this connection must be held to include the whole coast line of Palestine and Syria, together with that of much of the southern provinces of Asia Minor. There are few important points in this long and irregular seaboard which did not at one time or another come within the vision of the squadron ships or the seaplanes which they carried. It is giving the Turks no new information to tell them—what their coastguards fully realised—that many strange ships flying Allied flags were continually prying into waters off districts far removed from the scene of military operations, or to mention that some of these ships despatched flying machines with airmen who sometimes dropped bombs and nearly always brought back valuable photographs. Such information would lack the charm of novelty, but that quality would be particularly lacking with reference to visits paid along the Syrian coasts and along the few miles which extend just to the west of Alexandretta. Here, during 1916, hardly a week, certainly not a month, passed unmarked by one or more of these swoops of what I may call the carrier hawks.

To the armchair strategist who had built his hopes on some big move in this quarter, these spasmodic attacks from seawards, if he happened to hear of them—which is doubtful, since scarcely a word about them ever got into the newspapers—might well have appeared to be a policy of pinpricks. But they were not; and a glance at the map will show that they were not. It is all a matter of lines of communication.

The general appearance of the map of the Eastern Mediterranean should be familiar enough. It takes very roughly an oblong shape, with the Suez Canal, leading from Port Said to the Red Sea, down at the right-hand lower corner, Egypt along the bottom, Palestine and Syria to the right, and Asia Minor along the top. The island of Cyprus, a little above the half-way line, looks rather as though it had slipped away from the land in the right-hand top corner and left the space known as the Gulf of Alexandretta. The majority of the larger towns dotted along the coast and a little way inland, have names which inevitably carry one's mind back to far distant history lessons, biblical, classical or English.

Let us note some of them in order, beginning

ASKALON.

THE LIE OF THE LAND

with Alexandria, where the ancient lighthouse was one of the seven wonders of the world. Skirting the Nile Delta to the east, we come to Port Said, a mushroom growth of very modern times, and then turn northwards to Gaza, not quite on the coast, Askalon, where dwelt the daughters of the Philistines and where the Crusaders fought, and Jaffa, which used to be Joppa, a port, though never a good enough one to rival Haifa, about eighty miles higher up.

TYRE.

Across the bay from Haifa is Acre, another crusading centre and the farthest point reached by Napoleon in 1799, and to the north again are Tyre and Sidon, disguised now as Sûr and Saida. Next there is Beirut, which is about half-way up, and Tripoli—not the African one—about level with the south of Cyprus. Then comes Latakia, of which most smokers know at least the name, Antioch Bay, with Antioch lying up the river a few miles inland, and Alexandretta. From here we go west and pass few recognisable coast towns, though Adana and

Tarsas, of the New Testament, are not far off. For the moment that is as far as we need go.

We are not immediately concerned with the historical aspect of these places, though it has its fascinations (I remember for instance, the thrill which I learned that Tyre and Sidon were after all real towns and still in existence). What concerns us is that a railway running more or less parallel with the coast has touched many of them directly or with branch lines, and brought others appreciably nearer to the pale of civilisation, which at the time of which I am writing was another name for war.

A short time before the war began there were three separate systems of railway running over this country.

The first system, a small stretch of fifty-four miles, ran from Jaffa, inland to Jerusalem. In tourist days it was supported chiefly by visitors to the Holy City, and it was none too careful of their comfort. It had, indeed, the reputation of being one of the two worst managed railways in the world. Its partner in this unenviable celebrity was the Hedjaz railway, of which I shall have more to say later.

The second system—I describe it thus because, though different companies were involved, there was a definite connection—linked Haifa, Beirut and Tripoli with a line running from Aleppo, which lies to the east of the Bay of Antioch, through Damascus to Medina in Arabia. This line took a direction more or less parallel with the coast, north to south, that is to say, with a slight tendency westwards, till it reached Maan, where it went away about south-south east, and parallel to the Red Sea. The three Syrian ports of Haifa, Beirut and Tripoli were thus brought into touch with the main artery of railway, though they had and still have no direct connection by rail one with another. This railway, then, so far as Syria was concerned, consisted of a single line running

SIDON.

from Aleppo down to Maan, passing west of the Dead Sea, and it had three branches to the coast, one to Tripoli, one to Beirut, and one to Haifa. It can be pictured as a very old comb with only three prongs, all of them rather bent and battered. Lying astride the middle tooth connecting Beirut with Damascus, are the Lebanon Mountains. The southern one, running in from Haifa, divided what were in New Testament days the provinces of Galilee on the north and Samaria on the south.

The third system was the line which, when

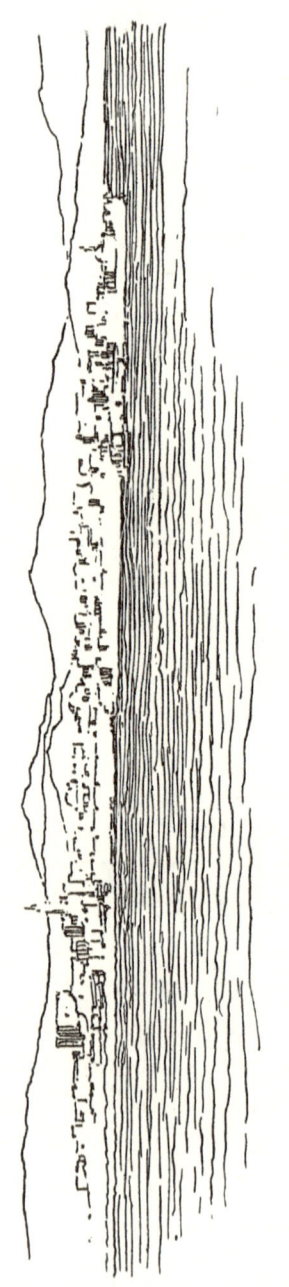

ACRE.

completed, was intended to link Constantinople with Bagdad—the famous Bagdad Railway. Of the places in our list of names it connected Tarsus and Adana, and running eastwards crossed the Chikaldere Bridge. Its route, as planned, came into contact with the second system near Aleppo, and ran thence in its long journey towards the Persian Gulf. At the outbreak of hostilities a considerable portion of this line had been laid, but there were two important breaks in it, and both of them formed gaps in the link with Syria. They were occasioned by the difficulties of tunnelling the Taurus range of mountains, and one of the first things the Germans did, even before Turkey declared herself an ally, was to send engineers and concentrate labour on the series of cuttings and borings in these two gaps. One of these gaps lay to the north-west of Adana, and the series was known collectively as the Taurus tunnel. The other series of tunnels to the north of Aleppo was known as the Amanus.

The directing of German attention on these engineering works at a time when Turkey still maintained her neutrality pointed to a moment in the future when that attitude, never particularly convincing, would be abandoned, and the moment was not long delayed. When Turkey declared war the anxiety of the Germans to complete the two tunnels was capable of a definite explanation other than that of a desire to reach Bagdad by railway. It appeared unmistakably as a double threat aimed at India and Egypt. Not only was work on the Bagdad Railway pushed forward with feverish energy, but work already in hand with the object of linking up all three systems received the stimulus of German encouragement. It is not our business to look very closely into the engineering details. It is enough to say that when the R.N.A.S. came to Port Said, the threat to Egypt had assumed definite proportions and Turkey was confessedly preparing for the great invasion which was to secure for Germany the Suez Canal, together with Egypt and most of the north of Africa. It became the job of the seaplanes to discover, so far as observation from the air could discover it, exactly what steps were being taken to bring off this splendid stroke.

The squadron came into being at Port Said at the beginning of 1916, and the Turkish invasion which was then, in prospect was not the first which had been attempted. The earliest effort had been made in December and January, 1914–15, very soon after Turkey's entry into the war. It was a very plucky effort, but excepting that it brought the Turks over the frontier and left them in occupation of various desert posts on the Egyptian side, it added little to their hope of conquest, and was from the military standpoint a complete failure. Their railhead—note this—was at Beersheba, and everything had to be brought from there by camel transport. Water was carried 180 miles; 15,000 men were engaged in the attack; a further 15,000 were held in reserve, about two days' march behind.

In the attack on the Suez Canal, one pontoon succeeded in crossing it, but it had no living occupant, and very few of the 15,000 in the front line survived. A great many died of thirst or starvation before they could regain their base. Had we been able to push forward then, we might have turned the defeat into something far more disastrous, but at that time we were no better able than the Turks to surmount the difficulties of the desert, and we had to stay where we were. There a strong defensive line was prepared, extending roughly from five to ten miles east of the canal, and behind this we waited with some calmness for the big offensive which Germany began to advertise in the course of 1915.

At the first attempt the Turks, as I have said, had a railhead at Beersheba, and the map with which the British Army was working showed no traces of it. It was an excellent map as regards geographical features—no less a person than Lord Kitchener himself had a hand in the making of it—but it was out of date as regards railways, for Lord Kitchener, when he worked on it, was a lieutenant in the Royal Engineers. It was prepared, that is to say, between 1878 and 1881. Revisions had been made, but still the only railways shown were the Jaffa to Jerusalem line and our three-toothed comb system of pre-war days.

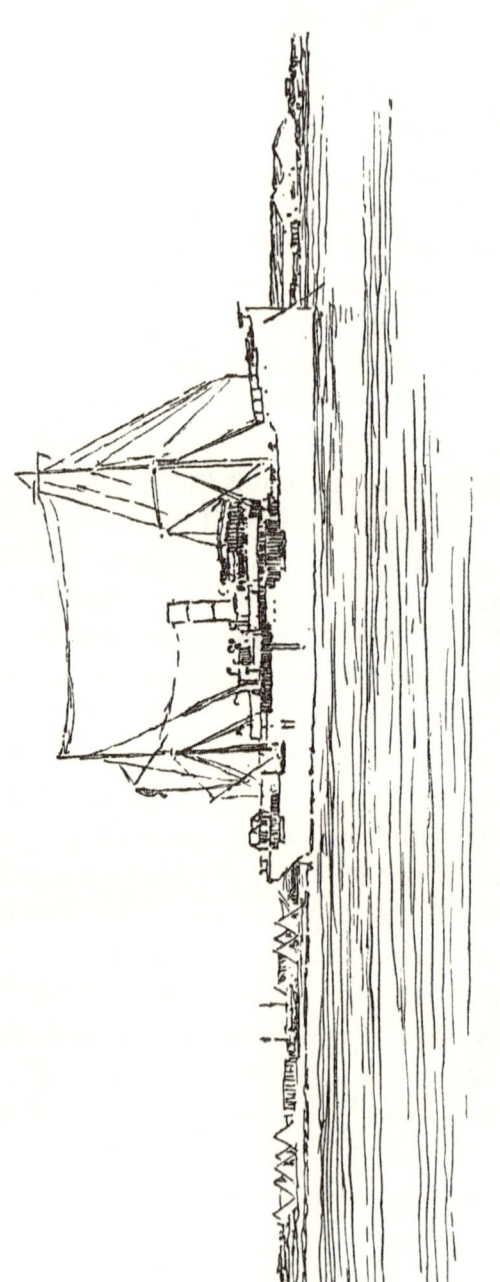

H.M.S. ANNE.

THE LIE OF THE LAND

In 1916 a new edition of this map appeared with approximate markings of the new railways, which were chiefly found south of Haifa, in the old districts, that is to say, of Samaria and Judea. A line at an average distance of about ten or twelve miles from the coast had been built north and south of Jaffa, crossing the Jaffa-Jerusalem line. The southern end of this turned inwards to Beersheba about thirty miles inland, and the northern end proceeded circuitously, also inland, towards the Haifa prong of the comb. Many branches appeared later, but at first the bare skeleton I have described was the sum total. The trace of this new line on the map was made from photographs and sketches taken from seaplanes.

Some of these seaplanes were French. The origin of the East Indies and Egypt Squadron is, in fact, to be found in the Anne, a ship which was captured at the beginning of the war, and was for some time lying idle at Alexandria. In the autumn of 1915, having been fitted out as a seaplane carrier, she was lent to the French to assist in their patrol work along the shores of Syria and Asia Minor. French machines were used and French pilots flew them, but the observers included several English military officers, one of whom, Major Fletcher, killed later in another war area, carried out much brilliant aerial reconnaissance in the early stages of the Turkish campaign against Egypt. The Anne remained under the command of our Allies till the spring of 1916. Meanwhile the Ben-my-Chree came to Egypt from the Dardanelles. This was in January, 1916, and in the course of the next few months she was joined by the Anne and the Raven, another prize, as we have seen, fitted out for the purpose. Thus was formed the squadron with its complement of three ships.

Its work, as I have indicated, consisted in harrying, by constant visits of inspection, the Turkish lines of communication. Our own front line was slowly but steadily on the move during 1916, but in the early days it was very close to the Suez Canal, and large tracts of desert east of Port Said were flooded as a defensive measure. One of the duties of the seaplanes was a periodical inspection of these inundations to note whether the water was properly contained, or whether it had forced outlets for itself. There was nothing very interesting about this, but there were excitements enough elsewhere. The railway builders had to be kept under close observation. The land machines of the R.F.C. took their share here, but much of the new line was beyond their reach, though it was well within that of the R.N.A.S. Numbers of valuable photographs were taken of the line under construction, not only in the approach to Beersheba, but also south of it to Auja, where it came to an end, though it was intended to go farther. It became intensely interesting to follow with the aid of these photographs the progress of the working parties, and to estimate the present condition and final direction of the permanent way. Wide deviation loops with a light "decauville" track of rails would show, for instance, that a substantial bridge was being built over some deep water course; the farthest point ahead would be marked by a roughly made embankment scarcely noticeable, except that there were shadows; behind these sections were those nearer completion with the sleepers laid in position, and behind these again appeared the newly finished parts with the rails laid, and in many cases covered with sand to shield them from aerial observation.

Not all the Turks were engaged on the railway, however. Scattered over the desert on the Sinai front—the "wilderness" of the Israelites—were camps and military posts, over many of which the seaplanes kept close watch.

Another long list of places would be tedious, and I omit it with the remark that, if written, it would contain many which are now familiar, because later they fell into our hands. But this happened eighteen months or more after the R.N.A.S. surveyed and photographed them.

Bombs were dropped too, and the bomb dropping was not all on our side. The Anne was attacked by German aeroplanes off Gaza on 16 April, 1916, when she was standing by for some of her seaplanes which were making a reconnaissance of a desert camp inland. Air raids were made on

GAZA.

THE LIE OF THE LAND

Port Said, notably on 21 May; and on 27 May the Ben-my-Chree returning from Haifa was attacked off El Arish by a Hun machine which dropped four bombs very near her and made off after half an hour's peppering from the ship's 12-pounder, 3-pounders, pom-pom and Lewis guns.

There was spotting also in these early days; on 18 May, for instance, when El Arish and neighbouring camps were bombarded and badly knocked about by two monitors and a sloop. Seaplanes from the Ben-my-Chree directed the fire and were responsible, among other things, for two direct hits which were scored on a hangar in the aerodrome.

This last scrap was described at some length in an Admiralty announcement, but the seaplanes carrying on their job behind the enemy front had a share in many others which passed unrecorded. The Turkish lines of communication were not entirely dependent on roads and railways. Stores of food and ammunition were frequently conveyed in small coasting ships, and when the seaplane carriers were out there was always a chance of encountering some such craft. One of these, described as a red schooner, gained a certain notoriety as a runner of contraband, and it ultimately fell to the lot of the Ben-my-Chree to put a stop to her sailings. It was in July, 1916, and the Ben-my-Chree, with a French destroyer as escort, was returning from a bombing attack on a railway station to the north-east of Gaza. About 4.30 in the afternoon the red schooner and two others were sighted, and our two ships immediately gave chase, while a seaplane was sent to head them off. The Ben-my-Chree engaged the leader, the red schooner, whose destruction had been specially requested. The crews abandoned the ships, and one of them was driven ashore in her efforts to avoid the attentions of the seaplane. As soon as the crew were away, the Ben-my-Chree sank two of them with her guns, and the French destroyer set the third on fire. The red schooner was evidently carrying ammunition, as a big explosion resulted from the first shell that struck her.

It is not my purpose here to give more than the most general idea of the nature of the work which was being done at this period. A detailed account of the squadron's activities would at best resolve itself into a string of dates and names. Looking back on them, one gets a rather confused mental picture of numerous expeditions, all of which justified themselves by the information contained in their reports. But the information was military, and military people have a liking for keeping that sort of thing to themselves, and turning it in due course to their own purposes. I don't know that we need grudge them this amusement. To the average person it matters very little that on such and such a day twenty-four bell tents were reported at Beit Where-is-it, or that 216 camels, not counting the one that strayed, were seen on the road from Dan

THE LIE OF THE LAND

to Beersheba. Even if I reveal to the average person the fact that on 29 February new trenches were being dug in square 070 G 12, it will be of very little value to him, because the trenches have long since been obliterated. But there were several outstanding operations which have an interest apart from such trivialities, and we'll see what can be made of them, without wounding the susceptibilities of our military hoarders of dull statistics.

IV

A ROUND OF VISITS

In view of my remarks on the curious standard of values set by the Military authorities on quite uninteresting items of intelligence, it is not an easy matter to select from among the squadron's records the most important of its operations, but judged by ordinary standards of magnitude, the scheme of which the attack on El Fule Junction formed a part must be given a prominent place. From first to last the proceedings occupied six days, 24 to 29 August, 1916, and all three seaplane carriers participated.

The general idea was two-fold. Firstly, a rapid series of effective attacks was to be delivered on enemy communications with the army on the Sinai front; and, secondly, a reconnaissance was to be carried out of all accessible points on the whole of the lines of approach to that front from Adana. This was a pretty big programme.

As we have seen, the Syrian railways have a main line running generally more or less parallel with the coast, but parts of it are some distance away, and at other places where it is not very far from

A ROUND OF VISITS

the sea there are intervening mountains which, doubtless, the engineers considered a sufficient safeguard against incursions. These operations of August, 1916, were among several which proved that if this was their idea, the engineers, German or Turkish, were mistaken.

The raid on El Fule Junction was the opening and principal item of the first part of the programme. This station, which lies on the southern-most prong of our original three-pronged-comb system,

MOUNT CARMEL AND HAIFA.

was the point from which the new line had been constructed in the direction of Beersheba and Sinai. It is between twenty and twenty-five miles inland from the nearest spot on the coast, and as the seaplane carriers would stand some miles out, it entailed a flight of at least thirty miles. But the thirty-mile route meant crossing mountains—the Carmel range, to be precise—and it was preferable to take a longer way in which there was not so much high ground to surmount. This way followed the railway which runs inland from Haifa

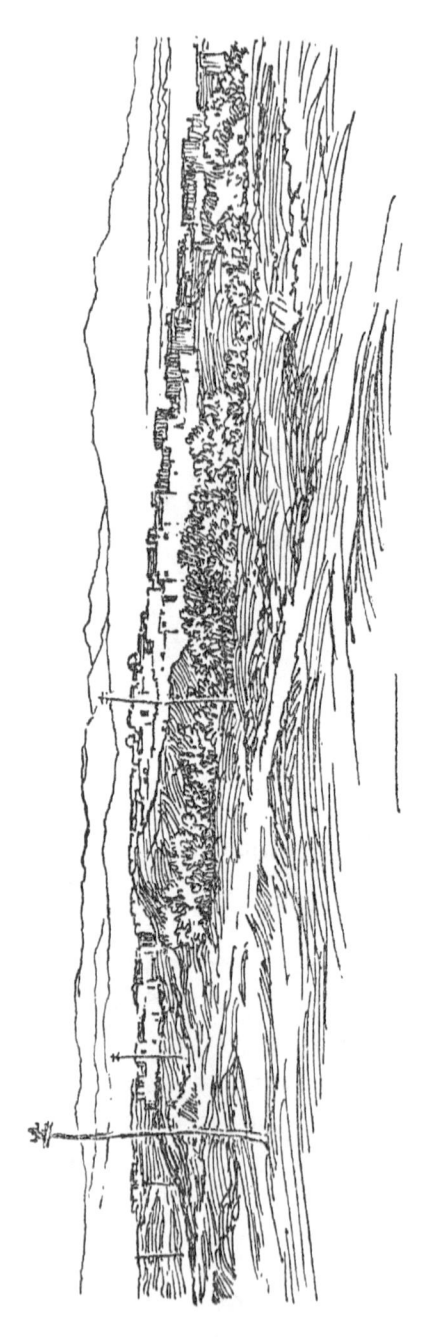

EL FULE.

A ROUND OF VISITS

up the valley between the heights of Carmel and Nazareth, and it added perhaps another ten miles to the seaplanes' journey.

Forty miles or so was not by any means an unusual distance for the Squadron's machines to travel over land; indeed, they often went a good deal farther than that, and no one thought very much about it. But it is worth while, all the same, to bear in mind that seaplanes are built for oversea work, and they can take off and " land " only on water. Serious engine trouble, during these inland flights, whether accidental or caused by enemy action, meant almost certainly that the machine must come down on solid earth with only its floats and no such wheeled undercarriage as aeroplanes have to land upon. The actual flying in these circumstances, apart from hostile attacks, was in many ways a more risky proceeding than a forty-mile flight straight across the German lines on the Western Front. There, it is true, the peril from artillery would be many times greater than in Syria, though Turkish anti-aircraft shooting was not at all bad at times, but in Syria, if the machine did get badly hit, it was a poor look-out for both pilot and observer, whereas in France there might be a fair chance of a safe, if difficult, landing.

This, however, by the way. El Fule Junction, distant forty miles or so inland, in the middle of the Plain of Esdraelon, or Armageddon, was the first objective, and just before dawn on 25 August, the three ships, having started independently from Port Said on the previous day, sighted one another off Haifa. The orders were explicit. The Ben-my-Chree's machines were deputed to bomb the rolling stock in the station at a height of 1,500 feet; to the Anne's machines were allotted the station buildings and stores at 1,700 feet; and those from the Raven had the task of smashing up the line leading from the Junction towards the south. Ten seaplanes were engaged, four single-seaters, and six double, and the special duty of the single-seaters, having dropped their bombs, was to act as a guard to the other machines in case any hostile aircraft was encountered. This precaution, however, proved to be unnecessary. On conclusion of the attack, each machine was to carry out a reconnaissance to supplement information gathered during the last visit to the district, which had been paid about a week previously.

The ten machines were hoisted out about 5.30 a.m., and a double-seater, distinguished by a fin painted red, which was to lead, rose from the water and flew in a circle round the three ships. This was the signal for the remainder to take station in the prearranged formation of " starboard quarter-line." One by one the seaplanes climbed into the air and proceeded inland.

The course followed had been selected, as I have shown, because it obviated some of the difficulties of airmanship in heavy sea machines operating over land. The country was comparatively low-lying, but there were numerous little hills—one of them the traditional scene of Elijah's defeat of the prophets of Baal—and many of these were entrenched and manned as a protection against possible attempts at invasion from seawards. The machine on the extreme left in the line of attack kept well to the far side of the valley route, so as to allow space for the extreme right to avoid the higher ground of the Carmel range, and this machine, the leading one, alone seems to have reached the objective without coming under rifle fire. At least, if there was any rifle fire, the observer didn't think it worth mentioning in his report. None of them, at any rate, could rise to an altitude which gave them immunity from this kind of annoyance, and those on the right in particular—the three machines from the Raven—were subjected to both horizontal and plunging fire from several enemy posts scattered about the Carmel ridge. Rifle fire was also encountered from Tubaun, a fortified camp about half-way on the journey. The ten machines, however, duly arrived at their several destinations.

At the Junction, news of their approach had preceded them. This became evident when a train, which had been sighted in the distance standing in the station, got up steam and hurriedly made off along the line leading south. Here, three miles away, it encountered the flight from the Raven, and the Raven flight made the most of its opportunities. The first machine damaged the track with its first bomb at 1,000 feet, and, coming down a couple of hundred or so, got another plumb on the rails and a third on the end coach of the train. This was the single-seater.

Of the two double-seaters, the first got one direct hit on the line and one on the embankment, and the other struck the embankment with two bombs five yards and fifteen yards to the rear of the train. This not only brought the escaping rolling stock to a standstill, but caused, as was revealed later, a definite and lengthy suspension traffic in that direction. Having thus successfully carried out their part of the programme, these three machines braved once again the rifle fire of Tubaun, which had grown rather more alert now that the camp had been awakened to the fact that something was happening, and regained their ship.

Meanwhile at the Junction itself, the remaining seven machines were having, and giving, a hot time. The reconnaissance of the previous week had encountered no anti-aircraft fire, but the Turks had been busy in the interval and had got two guns in position between the town and the railway station. These opened on the machines as soon as they got within range, and their efforts were supplemented by rifle and machine gun.

All the seaplanes were hit, but not seriously enough to hinder them, and the orders were carried out. Station buildings, store sheds, permanent way, rolling stock, tents, were effectively bombed, but the most spectacular display was that of the red-finned seaplane, which placed a 65 pounder very neatly between two trains standing parallel to one another on a siding at one end of the Junction. This wrecked both trains and damaged the line.

At the conclusion of the attack smoke rendered observation difficult, but it could not hide the traces of destruction. Moreover, it told its own tale. The railway buildings were burning at four different places. The seven machines left it at that and returned to breakfast, though not without receiving a few parting shots from Tubaun and elsewhere.

And even then the game was not yet finished. The slower ships, the Anne and the Raven, were sent on down the coast to the next rendezvous, which was Askalon, where they were to foregather again in the afternoon, but the Ben-my-Chree stayed behind and sent out two machines for a "look-see." They wanted to make quite sure that the observers of the early morning had not been carried away by the exhilaration of the moment, and seen what should have been, rather than what was. Also they wanted to drop a few more bombs. The first of these two machines went back to the Junction, knocked some more rolling stock about, and sailed off down the line to inspect the work of the Raven flight. It was even as had been

HAIFA.

A ROUND OF VISITS

reported. There were the battered and charred remains of the hindmost coach, and hard by there were uprooted rails and many breaches in the embankment. The machine had been hit three times when it arrived back again with its confirmatory reports. The second machine was not so successful. It provided the only example on this particular morning of that cussedness which is the bugbear of pilots and the bane of engineer officers. All the other machines had done their work well, and this one had already made one quite respectable flight. Perhaps it thought it deserved a rest; perhaps it never thought at all, though seaplane engines do at times seem to be able to think in a perverse sort of way; perhaps it had just dropped off to sleep and wasn't properly awake again. Anyhow, it put up a very sulky, sluggish performance. It refused to rise above 800 feet, and its pilot had to give up trying to urge it much beyond Haifa. There the observer made some useful notes of small camps before they turned round for home.

Curiously enough, on two other occasions, the air of Haifa, had, for some unexplained reason, a similarly depressing effect upon the Squadron's seaplanes. To one of these occasions I shall refer again later. The other was in March, 1917, when a single-seater, which had been flying over El Fule and its railways and was on the way back, suddenly dropped its "revs." to a speed at which the pilot could only with the greatest difficulty reach the sea.

Coming over the town of Haifa, he was scarcely 70 feet above the roofs, and he declared that he could tell from the smell of the smoke he passed through that the inhabitants were still enjoying a plentiful meat diet. This information was not forwarded to G.H.Q. for several reasons, one was that the pilot, having brought his machine down safely in the harbour a mile from the town, and passed an anxious half-hour or so before he was pluckily picked up by a French destroyer, didn't think of mentioning it until some days later, and then, somehow, no one believed it.

Having completed her last look round at Haifa, the Ben-my-Chree steamed south to catch up the other two ships. She was not to rejoin without an adventure. Just off Jaffa, two dhows were encountered. One of these, containing twenty soldiers in khaki, who were observed to take to the water and scramble ashore, was sunk by a French destroyer which was escorting the seaplane carrier. The smaller of the two surrendered, and in due course was brought to port by the Ben-my-Chree. Her crew of five were made prisoners, and stated under examination on oath that both dhows were engaged in victualling the Sinai troops. The history of this second dhow is worth following up. For some months it lay idle at the island, where also was lying idle a spare engine for a motor lorry. It occurred to some ingenious person that useful results could be achieved by amalgamation.

A ROUND OF VISITS

So the captured dhow was shortened and fitted with a new stern and a good many new timbers, and the lorry engine, supplied with a shaft and a suitable propeller, was put into place amidships. This took some time, but in the course of events a most serviceable motor boat emerged to begin a new series of trips which proved it to be one of the fastest and most efficient launches for its size in the harbour. It was still employed in revictualling, and the Turkish army, from all accounts, would have been glad to share the rations it carried.

In the afternoon of 25 August, the Ben-my-Chree overtook her companions off Askalon, whence attacks were to be made on the camp at Burier and the railway viaduct over the Wadi el Hesy, and reconnaissances were to be carried out on the line over the country in the direction of Beersheba. The bridge over the Wadi el Hesy was about fifteen miles from the coast, and the important camp of Burier was about half that distance on the way south of the bridge objective. The permanent track followed a very wandering course, which took it in places over twenty miles from the coast. The general direction of it was by this time fairly well known, though there were numerous corrections in detail to be made in the maps. In this particular section it passed among hills which necessitated many minor deviations and several considerable bridges. The positions of them were, in some cases, only approximately known, and part of the work of the reconnaissance machines was to make what contribution they could towards definite knowledge on the subject. The reconnaissance programme also included Gaza, which is about as far south of the mouth of the Wadi el Hesy as Askalon is north of it. It will be seen then that the work of this August afternoon of 1916 entailed making a rapid survey of a very important part of the battlefield which became famous when General Allenby made his advance in November and December, 1917. It was here that the left flank of the British Army took up its position after the fall of Gaza.

The seaplane operations were successful, but they were not attended with the same brilliant success which covered their morning's efforts at El Fule.

The viaduct over the Wadi el Hesy was not itself destroyed—it is no easy thing to destroy solid masonry with seaplane bombs—but the railway embankment was badly damaged, and there was a direct hit on the line itself. All machines experienced very heavy fire from guns and rifles at Burier Camp, and there were two forced landings, one machine, a single-seater, failed to return; the other, having dropped its bombs at random to lighten its load, managed to make the coast off Gaza.

After this the ships parted company. The Ben-my-Chree and the Raven went up north on various quests, which we will follow later. The Anne remained off Askalon and Gaza in the hope of picking up the missing pilot, who might conceivably have been able to make his way to the coast in the night, and attract her attention by light signals. Nothing came of it, however, and news was received later that his machine had crashed and that he had been killed.

Early in the morning of 26 August the Anne moved to a scene of operations slightly to the north. Her orders were to bomb Ramleh and Lud,

RUINS AT RAMLEH.

to bomb the railway at Tul Keram and to survey the line thence inland to Nablus, or, more familiarly, Shechem. These orders involved much of the country through which ran the right flank of the Turkish front line at the conclusion of the British advance at the end of 1917, and the beginning of 1918, when Ramleh, Ludd, Jerusalem and

Jericho had all been taken. Ramleh and Ludd are on the railway close to the point where the Jaffa–Jerusalem line came in contact with the extension which linked Beersheba with our three-toothed-comb system. It was at Ramleh that the Turks had their principal aerodrome during 1917, and there was a very considerable one there in 1916, when the town was a great deal farther behind their front. After the retreat from Gaza in 1917, Tul Keram became the resting-place of the fugitive airmen.

The Turks and Syrians of this particular district —Judea and Samaria of the New Testament—had thus plenty of opportunities of studying aviation. The period during which these studies were possible covered many months, but perhaps the most instructive single occasion was that of 25 June, 1917, when seaplanes from H.M.S. Empress bombed Tul Keram, while the R.F.C. engaged the attention of hostile machines at Ramleh. This was a rather neat piece of combined work of which the success was largely due to the Empress, who, having only recently joined the station to fill the vacancy caused by the loss of the Ben-my-Chree earlier in the year, was navigating off a strange coast, and was yet able to reach her position at precisely the prearranged moment.

The Anne's operations of 26 August, 1916, were more limited in the offensive scope, as there was only one machine available, but this made two flights and carried its investigations considerably

A ROUND OF VISITS

farther inland. Nablus is some thirty miles or so in a straight line from the coast, but the railway takes a very winding course, and nothing resembling a straight line of flight could afford the information relating to it which was the object of the visit. The reconnaissances revealed much engineering activity. A light railway, whose existence, had hitherto been unsuspected, trailed away northward from Tul Keram for the purpose,

TUL KERAM.

as was subsequently discovered, of getting timber from an oak forest which lies four or five miles inland from Caesarea. Observations farther on discovered many serpentine curves of railed track and stations under construction in the neighbourhood of Samaria, where the line turned northward and eventually led to the spot, south of El Fule, which the Raven's machines had damaged on the previous day. This flight took place in the morning. The flight over Ramleh and Ludd followed in the afternoon, and the Anne then packed up and returned to the base.

Meanwhile the Ben-my-Chree and the Raven had sailed north. The Ben-my-Chree's destination was the Nahr el Kebir, a river mouth some fifteen miles north of Tripoli, whence her machines were to make a reconnaissance of the railway running from Tripoli to Homs, the topmost tooth of our comb. The ship arrived early in the morning of the 26th, and two double-seater machines were hoisted out. The first failed to gain a greater height than 1,000 feet, and a reconnaissance of the coast defences only was effected. The second machine, however, amply compensated for the deficiency of the first, and put up a really good exhibition. The weather conditions were exceedingly unfavourable. A strong wind blowing down from the Lebanon Mountains, which lie to the south, made the flying extraordinarily "bumpy," and there were clouds at 1,500 feet which took thirty minutes to climb and obscured the land during the greater part of the outward journey. In spite of these handicaps, a straight course was steered by compass, and Homs station, forty-five miles away, was made dead ahead.

One of the objects of this flight was to test the truth of a report that the rails on this stretch of line had been removed for use elsewhere, and luckily the atmospheric conditions, which on the

A ROUND OF VISITS

way out had baffled the observer, had so improved on the return journey that important notes and photographs could be taken. The flight lasted two hours, and in the meantime a single-seater had been out to bomb a camp at Tel Keli. Similar atmospheric difficulties were experienced, but a fortunate gap in the clouds, just in the right place, enabled the pilot to deposit his load well within the required area.

The Ben-my-Chree then steamed to Ruad, and after coaling at Famagusta, in Cyprus, made her way on 29 August to Karatash Burnu, the headland immediately to the south of Adana, which will be remembered as one of the important towns on the piece of the railway between the Taurus and Amanus tunnels. Here three machines, one single-seater and two double, carried out the last item in her share of the combined programme. Four machines should have co-operated, but one of them jibbed at the last moment, and its pilot had to bring it back after half an hour's struggle.

As on all previous flights, each seaplane had the two-fold task of bombing and reconnaissance, and the courses covered radiated north-west, north and north-east of the headland off which the ship was lying. One machine inspected and bombed the small craft in the salt lake to the west of Karatash, and proceeded thence to the mouth of the river which runs down from Tarsus. The second made its way to Adana, where it bombed the railway station and the bridge over the river. The third looked out for lighters and dhows in the creeks and inlets of the coast of the Gulf of Alexandretta, and made a reconnaissance of the Imrik River which there enters the Gulf, and higher up is crossed by the railway at Chikaldere (our old friend).

The machine which went to Adana made on its return a second flight towards the Gulf of Alexandretta, looking out for submarines and paying particular attention to coast defences between the Gulf and Karatash Burnu. This completed the Ben-my-Chree's work, and she returned to Port Said.

The third ship of the squadron, the Raven, had gone farther afield. Her final task took her along the southern coast of Asia Minor well to the west of where the Ben-my-Chree had been operating. It is unnecessary to go into the details of her cruise. All that need be said is that she sent a machine to look at several suspected submarine bases, and they carried out their instructions. She was the last to return from the combined operation in which the ships had together covered 2,626 miles, and in which almost every accessible district of territory on the enemy lines of communication had been visited.

A day or two after the conclusion of this expedition the Turks paid us a return visit. As we have seen, it was not their first air raid, but this time it seemed that they had the Squadron's late efforts particularly in mind. One of their bombs hit the Raven and made a small hole in a wooden screen on one of her decks. She survived this injury.

V

THE COAST OF ARABIA

OPERATIONS in the Red Sea, which throughout 1916 occupied a good deal of the time of all three seaplane carriers, were always of longer duration than the quick raids up the Syrian coast, where the most lengthy of all was completed in seven days. In the first place, to reach any scene of actual hostilities, when once the Turks had been pushed away from the vicinity of the Suez Canal, meant a considerable voyage. It is some 300 miles or so from Port Said to the mouth of the Gulf of Suez, and from there it is upwards of 1,200 miles more to Aden, where the Squadron's ships operated on several occasions, voyages in these waters, therefore, had a rather more leisurely character than many of the others.

Theoretically every ship was ready for sea at a very few hours' notice, and for the shorter trips no more than a very few hours' notice was given. Practically, however, a trip which may last three weeks or a month does require more preparation than one which may be over well within three days. More food will be consumed for one thing, and the twelve hundred miles between Suez and Aden offered very few places where one could be certain of replenishing stores.

There was a difference, too, in the nature of the work. Nowhere in Arabia did the Turks present anything at all similar to the Sinai Front, as they called it when it was in Sinai and continued to call it when it was somewhere north of Jerusalem. The ground over which the seaplanes operated was generally occupied by small parties of Turks, nominally, but by no means literally, in control of native tribes, who, as it was discovered, were frequently ready to sever the somewhat frail bonds of their allegiance when the truth dawned upon them that it was possible to be attacked from the air.

There is a mixture of peoples in Syria, but no such bewildering mixture as is to be found in Arabia where all are Arabs. There are, to begin with, two visibly distinct groups: the Arabs of northern Arabia, who on the whole are a rather noble-looking people; and those of the south, who are far less attractive in appearance and lacking in most of the finer qualities which the others possess in a greater or less degree.

" The common idea of the Arab type "—I quote from Major-General Maitland's preface to Mr. G. Wyman Bury's entertaining volume, *The Land of Uz* —" is derived from picture books, and from travellers in Syria and Palestine, who unite in representing Arabs as tall bearded men, with clean-cut hawk-like faces.... These are the northern Arabs, perhaps the finest of the Semitic races. The Arabs of Southern Arabia are smaller, darker, coarser-featured, and nearly beardless. Their garments are so scanty, that when I was Resident at Aden, chiefs coming in from distant parts of the hinterland had sometimes to be provided with clothes before coming up to the Residency for their formal interview. All authorities agree that the Southern Arabs are nearly related by origin, as well as by subsequent intermarriage, to the Abyssinians. Yet, strange to say, it is this Egypto-African race who are the original and 'pure' Arabs, while the stately Semite of the north is 'Mustareb,' an 'instituted' or 'adscititious' Arab, one who is Arab by adoption and residence, rather than by descent. Nevertheless, it is maintained by all Arabs that both races are the descendants of Shem; the 'pure' Arabs, through the half or wholly mythical Kahtan or Joktan, Shem's great-great-grandson, and the Northern Arabs as the children of Ismail (Ishmael), Abraham's son by the slave girl Hagar. Even modern ethnographers apparently consider that there really was a common stock, existing far away back, perhaps long before the Sabean kingdom, from which both races are descended."

One may note here that travellers who have come into contact with Arabs, particularly those of the nomadic tribes, find that they endow with a strange air of reality these forefathers of theirs.

THE COAST OF ARABIA

As the Arabs of Arabia talk of Shem as they might talk of their own father or grandfather, so the Arabs of Sinai talk of Moses almost as though he were still alive, or at least, only lately dead, and of his wanderings as though they were the happenings of yesterday.

The inhabitants of Arabia, to whichever of these two broad racial groups they may be said to belong, fall into two other groups which were probably in existence in prehistoric times, though they are becoming more and more distinct through the march of civilisation—a very slow march in this country, whereof parts are quite unexplored. These groups, in the words of the authority already quoted, are " the settled inhabitants—cultivators, townsmen and traders, and the desert dwellers, or 'Bedou' (Bedouins), roving within certain limits and predatory."

The Bedouins, if we leave out of consideration altogether the more southern inhabitants, provide in themselves a pretty wide range of possibilities in the way of variety of political opinion. The chief man of a tribe may or may not regard himself as independent. In a general way he may acknowledge the suzerainty of some higher potentate whose influence may or may not be largely religious. He may or may not consider himself compelled to accept as binding upon him the policy which this higher potentate may see fit to adopt in respect of any yet higher authority. So that, when by such successive steps of vague responsibility one reaches the highest authority of all, as it was before the war, the Turkish Government, we are confronted with something which was a government only in name, and regarded its privileges and its obligations as completely fulfilled when it had graciously condescended to accept taxes farmed out and collected it cared not by what means or from what sources.

As in Syria, so in Arabia, though to not nearly the same extent, the war centred round the neighbourhood of the railways. But since Arabia possessed only the Hejaz railway, and that travelled less than half of the length of the country, there were many other districts in which the fighting was little if at all affected by it as a means of communication.

The Hejaz railway connects with our three-toothed comb portion of the Syrian system. The junction is at Deraa, where the southernmost tooth, running inland from Haifa through El Fule, joins the back of the comb. From Deraa it runs down east of the Red Sea to Maan, a town seventy miles or so north-east of Akaba on the Gulf of Akaba, which washes the eastern side of the triangle of land containing Mount Sinai. From Maan it takes a course following the line of the Red Sea, but never nearer to it than eighty miles from the coast. The terminus is at Medina. The full extent of the Hejaz line, which includes the back of our Syrian comb and starts near Damascus, is 820 miles. It is frequently called the Pilgrim line, and

THE COAST OF ARABIA

the original idea of its promoters was to provide means of transit for the Mecca pilgrims. Mecca is, however, some 220 miles due south of the terminus at Medina, and the pilgrims still reach it by way of the Red Sea port of Jedda, some forty miles away, and by numerous caravan routes.

Hejaz is one of the main districts into which Arabia is divided. It occupies the northern part; to the south of it is Asir, and between Asir and the Aden protectorate is Yemen. The Hejaz railway was constructed by German contractors, and for some time it was controlled by a German manager, who, some three years before the war began, handed over to the Turkish Government a line which compared favourably with other railways in the East. But Turkish laxity of management quickly earned for it the reputation which I have already mentioned, that of one of the two worst railways in the world, and conditions were not improved by the attitude of the people through whose country it passed.

The sharpest contrast exists between urban and rural life in Hejaz. The townsmen are among the best off, the Bedouins are among the worst off, in Arabia. Thanks to their detachment from the settled communities and to the poverty of the soil, the Hejaz Bedouins are exceptionally unproductive and uncommercial; and further, owing to temptation offered by the pilgrim traffic and also to the check on their natural free development so long imposed by the Porte and the Emir, they are of exceptionally predatory character, low morale, and disunited organisation. Though there are very large tribes there are no chiefs excepting the king himself. The tribes act less as units than anywhere else, and sub-sections frequently consider themselves free to go their own ways.

Before the war, the most important native personality was the Emir of Mecca, Hussein, an Arab who had had a European education. His power was really of a religious nature, but among Moslems that means that he had a good deal of secular power as well. He was quite friendly to England, and in the days of peace that was a point in his favour in the eyes of the Turkish Government which appointed him. When hostilities broke out, however, it was not quite so satisfactory. Hussein had the right of levying troops in his own country, and the Turks wanted to conscript them. He wouldn't allow it. The Turks wanted to extend the railway beyond Medina. The tribes had opposed this scheme from the first as likely to interfere with the revenue, considerable though variable, which they drew by exacting tribute from the Mecca pilgrims. Hussein continued the opposition to it because he knew that the extension was part of a big German scheme to use the Red Sea ports to mine the coasts, and as bases for insurrections to be fomented on the opposite shores of Africa and in Abyssinia and Somaliland.

THE COAST OF ARABIA

This attitude on the part of Hussein continued until the beginning of 1916. He was unfriendly to the Turks, but not strictly hostile. He was, in fact, waiting for a suitable moment to break with them. Similarly, the Arab ruler of Yemen was not by any means a wholehearted supporter of Turkish domination. He allowed Turkish garrisons in two of the principal towns, but his allegiance depended very much on his estimate of the Turkish commanding officer. Thus he disapproved of the attack on Aden in 1915 as an infringement of his prerogative, but later in the year he was once more on amicable terms with his Government.

At the beginning of 1916 there was a small mixed force of Turks and Arabs opposed to the British at Aden, and the R.N.A.S. began its work in these waters with a number of reconnaissance and bombing flights from the Raven on 3 March. These flights achieved excellent material results. Many direct hits were scored and good photographs were taken, developed and printed in spite of the extreme difficulty of having to use water which it was impossible to keep cool.

But far more important was the moral effect. The Arabs had never seen flying machines. Not only the civilian inhabitants of the villages and surrounding country, but also the native fighting men of the composite force, fled precipitately at their approach as from some invention of the devil. Neither by example nor by precept could their Turkish allies do anything to check the stampede. The Turks tried both expedients without avail. They stood their ground, and defended their positions with rifle fire. Had their fire been more accurate it might possibly have been more convincing. As it was, though the seaplanes flew low, none of them was hit, and the Arabs were not persuaded that any known weapon was effective against the strange menace. Deeds having failed, the Turks tried words. They told the Arabs that they would invent a device to stop the seaplanes dropping bombs and make them stand still. The Arabs may or may not have believed in the ability of the Turks to fulfil this promise, but it did not affect their immediate intentions. They were not prepared to take present risks on the chance of a future miracle, and they left the inventors to do their inventing unassisted.

A month later, in June, the Ben-my-Chree was at Aden. The position on land was that the Turkish troops engaged with ours in the neighbourhood of the town were finding the greatest difficulty in securing reinforcements and stores of ammunition. For supplies they had to rely on local produce, and to depend generally on the friendship of native chiefs. They had at the same time to maintain a show of authority for fear of losing their dominion over the Arabs altogether. This state of affairs offered us a particularly opportune moment for harrying them, but the intense heat

THE COAST OF ARABIA

rendered infantry action out of the question, and two methods of offence alone remained possible. One of these was the bombardment of coastal districts from the sea, and the other was attack from the air. In the circumstances it became the task of the Ben-my-Chree to deal with the towns in the Lahej Delta, which lies to the north of Aden.

The ship approached Aden at dawn on 7 June, and while she was still some distance off a seaplane was sent inland to make a reconnaissance of the Delta before the Turks could possibly have knowledge of her approach. This and a subsequent reconnaissance revealed the most suitable objectives, and during the next five days an almost continual offensive was maintained. It was achieved by the despatch twice each day of flights of machines with bombs, and evidence was quickly forthcoming that the Turks were not only bewildered by the unexpectedness of attacks which were constantly being directed at fresh spots, but completely deceived as to the extent and nature of the projected demonstration. Reports made it clear that the Turkish commander expected an attack by land to follow the attack from the air, and that he succeeded after desperate efforts in obtaining reinforcements. He also increased his transport by commandeering local camels.

Nothing further was heard of the invention which was to keep the seaplanes in a state of paralysis in mid-air, but a good deal was attempted in the way of aerial gunnery. Shell, rifle, and machine-gun fire was invariably opened, and the seaplanes were occasionally hit, but no serious damage was done in spite of the very low altitude at which, owing to atmospheric conditions, the machines were compelled to fly. One machine, for instance, which had taken out a large bomb, intended for an ammunition store in Lahej, could reach a height of only 300 feet above the town, and the bomb could not be dropped for fear of damage to the seaplane from the explosion. It was, however, discharged at a gun emplacement, which it demolished. In fact, so far from constituting a serious menace to the invaders, the improvised anti-aircraft defences became in Turkish hands a rather perilous weapon. Eight Turks were reported killed by one of their own shells, and one of the guns which had been put on the top of a palace fell through the roof. The commanding officer gave as an excuse for not succeeding in bringing down the seaplanes that the object of the attack was to make him exhaust his ammunition, which could not be replaced.

Throughout this period, as indeed during the majority of the Red Sea operations, flying was carried out in the face of extreme difficulties. Owing to the climate there was very little "lift" in the air, and the great heat caused the water in the radiators to boil and limited the time in the air or even prevented the machines from rising at all. One single-seater stopped dead, and the pilot had hur-

THE COAST OF ARABIA

riedly to drop his bombs to lighten the machine, and was then only just able to reach the harbour, where he came down in 2 feet of water.

After five days in the Aden district, the Ben-my-Chree steamed westwards and arrived off Perim Island early on 13 June. This island lies off the coast in the straits of Bab el Mandeb, the narrow outlet of the Red Sea where Arabia approaches nearest to Africa. Here she was to combine the two possible methods of offence already noted. There were Turkish camps close to the coast, and the Turks were trying to maintain as a line of communication the coastal road running north. This situation gave the Ben-my-Chree the opportunity not only of making extended seaplane reconnaissances inland, but also of bringing her own guns into action and using her machines for spotting.

This programme provided a sufficiently exciting day. The ship took up a position, north of the camps and 2,000 yards from the shore, which permitted of enfilade fire. The fire was indirect, since the camps were concealed by low hills, but as a preliminary a seaplane dropped two incendiary bombs whose smoke gave the guns their line. Several hits were registered, and these and the bombs from the seaplane silenced the gun which had been replying to the ship's bombardment. More effective shelling was experienced from a hillside battery. Two shells straddled the ship and shrapnel fell on the deck. Only one of these guns was silenced, and that only temporarily, but the Ben-my-Chree completed her task in this direction without mishap, and proceeded to another position. Here fresh camps were found and bombed, and while effective fire was maintained with the assistance of wireless direction from a seaplane all hostile shells fell at least 500 yards short.

This little scrap finished, the Ben-my-Chree moved on to Jedda, where she was to join in operations which are more familiar to the general public than the small engagement just mentioned. We are now dealing with the Hejaz again, and, as we have seen, Hussein, the most powerful Arab, had for some time been showing anti-Turkish sympathies. Both in Mecca and Medina he had reduced Turkish authority very low, and in Jedda his influence was superior if not supreme. But the Porte was maintaining its garrisons in spite of temporary interruption of railway communication, and under their protection the Ottoman officials held on. In May, 1916, Great Britain, who had been fully aware of his friendly attitude and were quietly helping him, openly stepped in. The Turks were getting supplies through Arabia to the Sinai front, and the British Fleet enforced a strict blockade of the Hejaz coast, thus closing the trade with India and the Soudan from which came corn and other necessaries. This, of course, cut off Arab supplies too, but that was taken into consideration and arranged for. Hussein then declared the independence of the Arabs, and assumed the title of King of the Arab Nation. This was on 5 June.

His earliest step was to invest Mecca, a move of the utmost significance in Mohammedan eyes, and the town was completely in his hands by 10 July. In the meantime British warships had been bombarding the Turkish defences surrounding Jedda, the port, it will be remembered, which gave access to the Moslem holy city to pilgrims from overseas.

It was on 15 June that the Ben-my-Chree arrived on the scene of operations, and in the evening three seaplanes, two single-seaters and one double-seater, were hoisted out to reconnoitre the garrison positions and to attack certain definite objectives not easily reached by gunfire from the ships. Photographs and observations showed that the Turks were in great force, and the two-seater, being bigger and slower than the other machines, was frequently hit. One bullet struck the pilot's foot, another nearly severed the elevator control, and others pierced the propeller. Further flights were arranged for the following day, but it was discovered that the town had surrendered. This information was sent to the Ben-my-Chree by the senior naval officer in a signal with the gratifying remark that " probably the seaplanes had decided the matter."

The Turks remained in possession of the Hejaz railway, with strongholds at Maan and Medina,

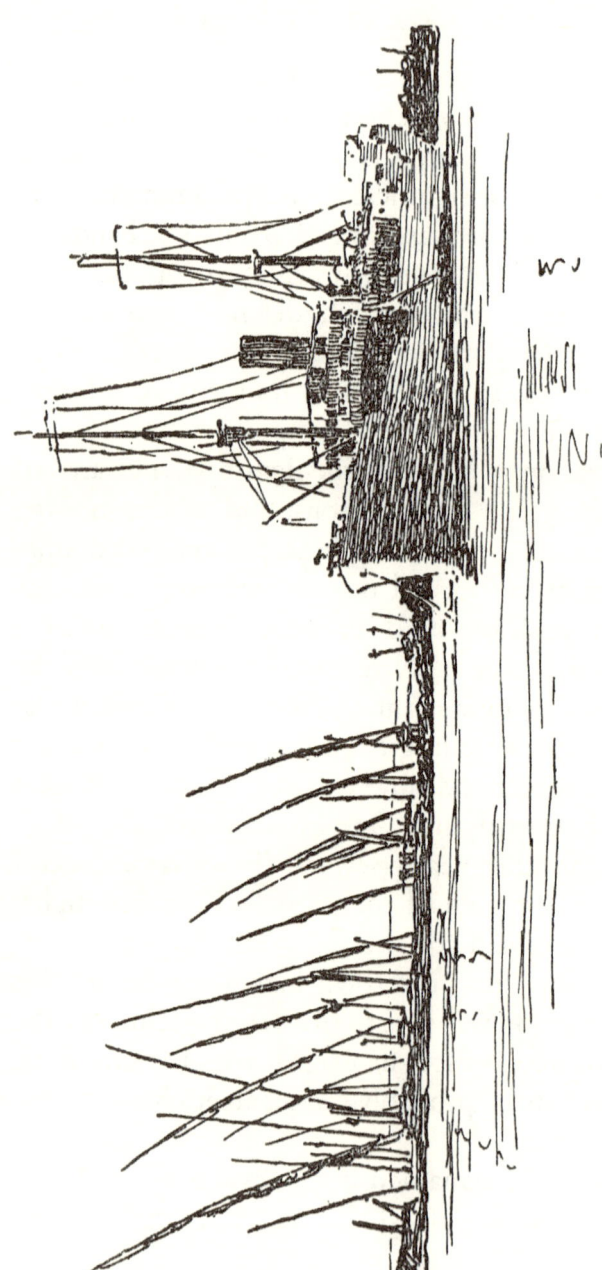

H.M.S. ANNE.

THE COAST OF ARABIA

and all their efforts were directed towards the recapture of Mecca. One of our first schemes to prevent this was to effect a landing in the Gulf of Akaba, push up to Maan, and cut off Turkish supplies by railway. As a preliminary to this move, the Raven, between 26 July and 7 August, carried out an elaborate reconnaissance of the town of Akaba and the country behind it. The information obtained there and at other places along the coast to the south, which was visited later during the same trip, showed that the Turks had strong garrisons in many of the seaboard towns. A trench system which was photographed near Akaba revealed a state of affairs of which we had no previous knowledge, and altogether the information was such as to make the landing scheme inadvisable. We therefore proceeded to reduce the coast towns, and one after another these came into the hands of the friendly Arabs. Jedda had fallen on 16 June; subsequently Akaba, Muweilah, Yembo and Rabegh, were taken; and finally Wej, the strongest of them, capitulated on January 28, 1917.

During September and October, 1916, the Anne was constantly at work long the Hejaz coast, and in December she was relieved by the Raven. Spotting for naval guns, bombing and reconnaissance were all carried out as the occasion demanded. The experience gained from the first Arabian flights from the Raven at Aden in March was also turned to good effect. Frequent exhibitions of airmanship were given with the sole object of impressing the natives. Machines flew at low altitudes—it was difficult in any case to rise to higher ones—over the towns with their curious mud-huts and their massive though futile mediæval-looking square forts surmounted by crenelated battlements. Arabs stared in wonder at the strange portents humming their messages in an unknown though not unintelligible language. Even the camels, shaken from their immemorial calm, did unexpected things, rose clumsily to their feet, or sat clumsily down. The Turks involuntarily relaxed the grip which had at the best never been anything but precarious. To the native mind, there was no question which master was the one to serve. It was not the one whose bullets seemed always to go wide of the flying mark or who talked vaguely of wonders coming, but which never came, to defeat the wonders already there; it was the Chief of their own race, who was friendly with the nation which wielded these dread weapons.

But although the Turks in Arabia never brought a seaplane down, their bullets did not always go wide. It was during the bombardment of Wej that the R.N.A.S. had its only Arabian flying casualty. One of the observers, Lieutenant Stewart, of the 7th Royal Scots and R.F.C., was hit by rifle fire while spotting for the gunners of one of H.M. battleships. Curiously enough—such are the fortunes of war—the machine had passed unscathed through fairly heavy fire and had, as it was supposed, drawn out of range, when the fatal shot, evidently a chance one, was received. The machine returned immediately to the ship, but the fleet surgeon pronounced life to be extinct. Lieutenant Stewart was buried at sea on the following day.

VI

ON DUTY AND OFF

ANYONE who goes to the trouble of sorting out the dates which I have jumbled into these very unchronological notes, and adding from his imagination a few more which I have left out altogether, will get the tolerably accurate impression that during 1916 at any rate, and during a great part of 1917, the three ships of the squadron were nearly always away. If they were not up the Syrian coast they were down the Red Sea—we generally talked of " up " and " down " in these connections—and operations followed one another so quickly that there was very little breathing space between whiles. But the base was there all the time with its permanent personnel, and the shifting population of flying men and mechanics, who, like Detective-Inspector Bucket, were here to-day, gone to-morrow and back again the next day, were ashore often enough and long enough to make the island a very pleasant and social spot when they were off duty.

As a place of residence, the island, like Port Said itself, could boast of few architectural attractions.

ON DUTY AND OFF

Three or four brick and plaster buildings, used chiefly as hangars and workshops, were supplemented by many more structures of wood. Some of them were full-sized huts of a standard pattern ; others were huts not of a standard pattern, though they were made of standard parts which the ingenuity of naval carpenters, with Arab assistance, had put together like a species of jig-saw puzzle that looked more or less right in spite of there being some pieces missing. Besides these, there were other wooden structures made out of that very useful perquisite of most air stations, the packing cases in which flying machines travel overseas. The stranger threading the maze of passages among them was reminded of nothing within ordinary knowledge, and imagined, perhaps, a resemblance to some desert island peopled by shipwrecked mariners.

Yet it was not without a scheme and a design, though there was a good deal of desert about it. In a vague sort of way, the buildings had a focal centre in the "quarter deck"—a patch of sand which had been at some distant date roughly cemented over. Perhaps originally the cement presented a continuous, square, level surface. It was not so in our time, but a few sandy depressions made it no less the quarter deck where the ratings would be mustered for various duties or for liberty, just as if there were a proper mast or a flag-staff, instead of an untidy pole with a more untidy yard, and a most untidy gaff. People at Navy House, with its spick and span masts and signal haulyards, used to stop the war every now and then and gaze over towards the island to laugh at our crazy flag-staff. But it flew the white ensign with the best of them.

It was in close proximity to this quarter deck and flag-staff that the experiments were made which proved finally that the island was a desert island.

R.N.A.S. ARCHITECTURE—I.

Guy, who came from somewhere in the Western wilds of America, instituted a scheme of afforestation. He planted two date palms and two banana trees. Deep pits were dug for them, and they were fed with water and rich meals of manure. Fences were erected round them to shield them from the sun and from the attentions of Guy's troupe of strolling live stock, consisting of a gazelle, a goat, and several kids. If you looked over the top of the fences you saw brown tufts like feather dusters made of matting. Guy used to take people to see them, or try to. Anyhow, he went frequently himself in order, I think, to make sure of being in at the death. He tended these pathetic three-foot trees as though they were children, but he did not put up the notice which stated that all ranks were forbidden to climb them.

R.N.A.S. ARCHITECTURE—2.

But although the island was barren of vegetation, it did not really resemble the desert island of romance. One might possibly mistake the R.N.A.S. personnel for shipwrecked mariners, but one could hardly so account for the Arab working party, which was ubiquitous during the hours of daylight. They were a picturesque crowd, these Arabs, from Big Ali who could do, and sometimes did, as much hard work as any ordinary navvy, down to the elderly couple of whom one spent his day watching the incinerator and feeding it with carefully selected morsels of rubbish, and the other walked round and round the water's edge gathering jetsam in a basket, and feeding himself with any particularly appetising tit-bits. Between these two extremes of native physique lay others of many sizes and hues, all of them answering to one or more of the three names, Ali, Abdul, or Mohammed, and all of them convinced that no job, however light, could be accomplished single-handed or without the accompaniment of a monotonous, droning incantation which contained, I am told, a reverent reference to Noah getting his charges into the Ark.

These gentlemen arrived by boats early in the morning, and we got up, or listened to others getting up (as the call of duty demanded), to receive them. To say that we listened to others getting up is to say no more than the truth. There was one officer, W.O.W., who invariably leapt from his bed, a home-made spring mattress of three-ply adequately supported at the edges, with a full-throated rendering of the naval reveille:

> Heave out, heave out, heave out!
> Lash up and stow, lash up and stow!
> Show a leg, show a leg, show a leg!
> Rise and shine, rise and shine!

While dressing he would give us snatches of songs recognisable rather by the words than by the tune, always opening the entertainment with one beginning, "Somewhere in the world the sun is shining." We used to refer to this as his weather report.

Fitful gusts of this cheery optimism of his would float down the length of the huts, and by degrees one would hear the occupants of the cabins awakening into life. These cabins were formed by partitions made from the invaluable packing cases, hastily fixed into position in the first place, but subjected according to the taste of the occupant to many and various subsequent processes of decorative or utilitarian treatment. The show pieces were, I think, K. E.'s and Guy's, each of whom had turned his quarters into a small museum of relics of his travels, set off with a finely conceived background of white and grey service paint, through which appeared irrepressible patches of creosote. This gave a mottled look to the white, but much of it was hidden under a careful arrangement of trophies celebrating the successful blandishments of local dealers in "antique" Soudanese weapons. There were also hangings of tent work embroidered with quaint figure-subjects taken from Egyptian tombs. These were the cabins *de luxe*. Another type was that which depended for artistic effect entirely on pictures of ladies cut from *La Vie Parisienne*, ladies who, if they wore clothes at all, wore garments eminently suited to a spot devoted to the intimacies of the toilet. Others, again, scorning silken dalliance, would contain the barest necessities for sleep

BEDOIR AND BOOTOIR.

ON DUTY AND OFF

and ablution. W.'s was the most austere of them all. Whether from a natural distaste of all comfort or from a desire to enjoy the greatest possible pleasure in leaving his couch at a very early hour in the morning, he would lie down and sleep most unmistakably on a hard shelf of bare planks, and such furniture as he possessed was stamped with the same ascetic simplicity. A box served the dual purpose of a chair and a chest of drawers, and half a dozen nails in the wall were an adequate wardrobe. If one suggested that there was not much protection from the sandy dust which at certain times of the year permeated every fold and crack and crevice, he would ask in reply why one supposed he paid Private Bedford. He objected on principle to what he described as "boudoirs." It was partly in defiance of this objection, and partly from the liking for a compact orderliness, that one of the military observers divided his cabin into two compartments. He called one his "bedoir" and the other his "bootoir."

Breakfast would be over by 8.30, and, the stand-easy from 10.30 to a quarter to eleven having thrown new strength into the late morning, the business of the day would show an appreciable slackening towards noon. From then on it would be difficult to keep up the semblance of full pressure. The sun in Egypt is at its hottest from about 1 till 3, and even at not quite its hottest, it is far from stimulating. Born Egyptians and transplanted

Englishmen frankly indulge in the siesta, and between noon and 4 p.m. slam their doors disconcertingly in the faces of strangers with continuous eight-hour-day ideas of business. Behind these forbidding doors one suspects that they sleep. It was not given to us to do this, but the reins were subconsciously held with a looser grip in the afternoon, and when lunch was over it did one no harm to lie down for half an hour somewhere in the shade, as the Arab working party would do to a man. It is too much to say that when the klaxon horn sounded at a quarter to two one rose as a giant refreshed, but one was certainly sufficiently vigorous to be glad of tea by four o'clock or thereabouts. After that, there would be liberty boats for shore parties.

After working hours there was not very much to do " on board " the island, and most of us went ashore. But there were one or two notable exceptions. One of these was L. C., who at one period of his connection with the Squadron hardly ever went into the town at all; and the other was P., whose head was always teeming with figures which appealed to him in every possible relation. At one time he would be merely working out the total financial outlay represented by the station with a view to discovering exactly what it cost to drop a bomb on a Turkish objective. At another time he would be contemplating chapters of a monumental work to be entitled, so we gathered, " Money for Nothing."

ON DUTY AND OFF

This at least is what we supposed would be the ultimate outcome of many hours spent in company with a turf guide, and many more hours devoted to the designing of an infallible system for use at Monte Carlo. There were other schemes, but none of them was capable of any very thorough test in Port Said, so there was not much to be gained by going there. He had, however, one other ambition, and that was to learn French. This took him shorewards perhaps two or three times a week. We used to look forward to his return, not so much to learn of his latest achievements in French or in the theory of high finance, as to hear his description of his teacher who, he told us in a whisper, was really a member of the Montenegrin Royal Family. "It's not known, of course. Incognito. But I have it on the best authority." There was a romantic glamour about that.

There was a touch of sadness about L. C.'s case. In the earlier days he had found a pair of bright eyes which seemed brighter when he approached. But after a while something happened. I think the bright eyes glowed brighter still elsewhere. At any rate, L. C. condemned himself to walks round the island. At first the walks were solitary. Then he bought a monkey, and the aching void was filled.

Transport by water was one of the perennial harassments of the Duty Officer. When the seaplane carriers were in port their boats might lend a hand, but when they were away there were times when we were reduced to one evil-looking, powerful, black tug, an enemy prisoner who still bore her pre-war name of Vulcaan, and, for lighter loads, a small craft with a motor attachment, which seldom did justice to its maker's reputation. To get this little incorrigible to start at all meant vigorous action like the grinding of an exaggerated coffee-mill, and when started she would go coughing away with a very palpable intention of selecting the most inconvenient spot for a breakdown. Many times her hapless crew drifted in mid-stream while the coffee-grinding process was alternated with a hasty vivisection of her organic parts; more than once they had to land at the wrong island, leaving her to be reclaimed by a rescue party while they bought or borrowed passages from casual passing Samaritans. But one didn't mind much. There was a war on, and this was all part of it.

More successful liberty trips, those, for instance, in the converted dhow captured by the Ben-my-Chree and fitted with a spare lorry engine, took us swimmingly down the fair way, rocking in our wake feluccas deep laden with grimy natives from the coaling depots—twenty or more of them in a boat incredibly propelled by a single pair of oars; rocked in our turn, in the spreading wash of incoming M.L.'s; past the many arched white front of Navy House; past the dry dock with its frequently changing occupants; past the triple-domed and colonnaded Canal Offices, almost the only building in Port Said which seemed in keeping with the Orient, though it was rather Italian than Arabic; past the moored tugs of the Canal Company with their white-ringed black funnels; past the destroyers, sloops and gunboats of all allied nationalities waiting their turn at escort duty; past the great splay-sided mass of the old French cruiser Jaures Guiberry; past the anchored units of the latest convoys; past, in short, all the fascinating life of

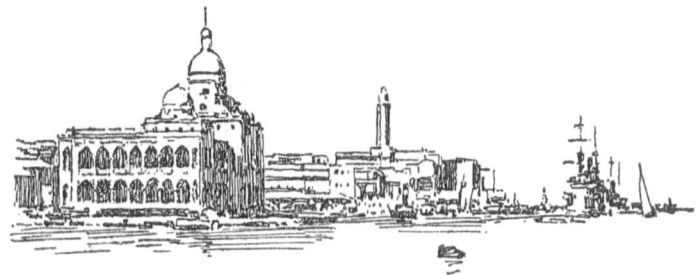

PORT SAID : THE SUEZ CANAL.

the biggest shipping port in the eastern Mediterranean, and on to the crowded and insignificant landing stage of the Customs House, with its shabby examination sheds, its police station, and its verandah containing the branch of the league for the suppression of the White Slave traffic.

Here one entered the town which, as our trip from the island progressed, had grown, unlike the taper waist of the fair one in Campbell's poem, large by degrees and hideously more. Seen from a distance, Port Said presents, as indeed do most distant towns, a picturesquely blended irregular group, but a feature or most such views is curiously lacking. Generally the sky line is broken with towers, spires, domes, minarets, or even factory chimneys; here the handsome octagonal column of the lighthouse provides almost the only line of relief from the horizontal. The nearer view dispels the picturesque, and one has the impression that the flat square edifices have been abandoned by their builders before the removal of the scaffolding. Nearer still, the scaffold poles rising from the kerb stand revealed as the supports of balconies which, hung with awnings and matting blinds, not only shade the shop-fronts and footpaths below, but offer unexpectedly cool sitting-places above.

At the gates of the Customs House one confronted the lures of the Rue de Commerce and, prominent among them, the signboard of Jim Irish, "well known to the British Army and Navy. Established 1882." Jim Irish—whence he got his name I know not—was an Arab contractor and marine-store dealer who was reputed to be worth five figures—six, perhaps—sterling, who could buy a pleasure launch at a war price and pay for it in hard cash running to nearly a thousand pounds, and who had mysteriously in the hollow of his hand every native labourer in the district. He was a wire puller, a man with many irons in the fire, one to be reckoned with; a man of whom it was well to be aware, if, emerging from the Customs House gates, one had business in Port Said and was not merely looking for first-hand impressions of the glamour of the East.

Not that this in its true sense was to be picked up in the Rue de Commerce or, indeed, anywhere in Port Said. One realised very quickly that the place was a port of call, that its small cosmopolitan group of residents lived upon the infinitely

COALING, PORT SAID.

larger and more cosmopolitan shifting population of tourists, shoregoing seamen, pursers in need of stores, and engineers in need of coal. Of the huge coaling traffic the casual visitor need see very little, though he would become unpleasantly conscious that it existed if his ship coaled during his stay, for the attendant lighters, with their swarms of very active basket carriers running up and down the narrow plank gangways and filling the air with the dust of their burdens and the shrill cadence of their songs, could remain unnoticed only by the deaf and the blind.

But other traffic was thrust into the face of every one who landed. The Rue de Commerce was lined with shops bearing names, showier, perhaps, but less weighty than that of Jim Irish, the names, Indian, Egyptian, Japanese, Arabic, Greek, even English, of dealers in ship's chandlery, clothing, books, and the innumerable list of wares one expects to find at a druggist's. With all of them one would discover a surprisingly wide range of goods apparently but little affected in quantity by the perils of oversea trade, though the prices were an unmistakable index of the continual rise of freight and insurance charges. The book-shops made interesting hunting grounds. On the handiest shelves were a wonderfully representative selection of the latest publications in fiction, and there was a considerable body of the more elderly classics. But it was in the high and out of the way shelves that one found surprises such as Victorian yellow-back novels and volumes with quaint wood-cut illustrations, derelicts from the days when booksellers, in this part of the world at least, knew nothing of the blessings of " sale or return " or of " remaindering."

But by far the most prominent among the shops were those which displayed curios. They stood open and ready to meet the hurried requirements of ship's passengers ashore for a few hours and anxious to carry away with them some reminder of their visit. And here, too, one looked almost in vain for local products. As with the groceries, books, chemicals and wearing apparel, the greater part was imported. Carved ivory, cloisonné ware, earthenware and porcelain, Damascened metal work, embroideries, silks and muslins from Japan; carved ivory of a cruder cut, trumpery brass and silver and printed fabrics from India; brass work

PORT SAID : THE MARKET.

enamelled or inlaid with silver and copper from Damascus. Egypt sent contributions in the form of metal work made in the bazaars of Cairo.

This was the Port Said which most people saw who paid their visit in war time, and it was through this that the shore parties from the island passed on their way to the Union Club or the Sporting Club, those gathering points which foreign influence,

NILE STREET, PORT SAID.

chiefly British or French, had formed to render the place habitable. The casual visitor, whether in uniform or not, passed through it, frequently not without paying toll, having run the gauntlet of the eager groups of salesmen standing in the doorways and out in the road. Indeed, one might pass through it many times before familiarity made one immune from their blandishments heralded by the mechanical invitation, " Look, Mr. Officer Captain. Come inside. Never mind buy anything. I show you. Very good, very cheap, very clean."

At the Union Club there were English billiard tables and English newspapers, the latest telegrams, and a pleasant verandah from which one could look down and study the headgear of the sauntering populace. At " Le Sporting " there was tennis, exhibiting many grades of excellence, ranging from the expert expositions of K. G., Scottie, Wallah, Pack, and Puck, the Half Blue, to the erratic displays of Slamby, who believed in hard hitting without regard to range or obstacles, and of Patterby, whose one idea was to make the game a social success with the aid of cheerful conversation; and thence down to the dignified posturings of Herbert, who played, in the words of Slamby, " like the hinder part of a stage elephant."

After tennis, perhaps, to fill in the interval of waiting for the return boat from the Customs House, one would sit at the tables in front of the Eastern Exchange Hotel, that hideous iron-

and-glass-fronted erection which is at the heart of things where the main streets intersect. Here, munching peanuts, taken warm at a penny a handful from the baskets of privileged hawkers, one could watch the light craft of Port Said high life sailing to and fro along Nile Street in the best of its holiday rig (purchased, perhaps, from " Au Printemps" at the opposite corner of the cross roads), carefully convoyed by the heavier tonnage of the maternal generation, which observed the artillery of glad eyes—sharp shooting rather than broadsides or salvoes—and took care that if there were occasional hits there were very few captures. Or one could watch the traffic in the doorway of the hotel itself, and speculate on the business which brought it in or out. There were those, for instance, who emerged looking merrier in mind and lighter in pocket; they had been taking cocktails at current rates at the bar. There were those who seemed to have done some ardent physical exercise; they had been driving bonzoline round the most sluggish billiard tables in Africa. There were those who drew their forefingers with a razor-like motion between their collar and the nape of their neck; they had merely been having their hair cut, and were trying to rescue the stubble which had slipped through the exiguous meshes of the wadding.

And so back again to Mess on the island, where a rather exceptional Greek did the catering for us

ON DUTY AND OFF

and stood the racket of our constant moans with wonderful imperturbability. Possibly he knew that there never was a mess without its grousers, and was shrewd enough to see that they did much to keep the standard up to the mark. One is liable to go easy in warm climates without constant reminders. Possibly he knew, though he never took advantage of it, that some of us had a quiet liking for him on account of a certain fine act of his which was done spontaneously without any sort of ostentation.

It was one Sunday morning. The business at the Base was proceeding at a quiet Sabbatarian pace when there was a sudden alarm. A gun was fired in the harbour, and we were aware that many thousand feet overhead, a small speck in the sky, there was an enemy aeroplane. Immediate orders were given to get out a single-seater, a quick-climbing seaplane, which had at least a sporting chance of coming near the intruder if only it would rise from the water in time. And that is exactly what it would not do. It was got afloat and started up, but 20 feet from the shore the propeller slackened down and stopped. Our Greek caterer, dressed in a quite smart-looking Sunday suit of blue serge, was nearest to it, and before one realised it, he was in up to his waist lending a hand. Few people of his nation—and I am not at all sure that the national question has very much to do with it—would have made that instant ungrudging sartorial sacrifice. And the pity was that it was useless. The Hun got away unpursued.

Our dinner-table conversation was, I think, very much the same as that of other flying messes. There was the regular undercurrent of "shop," but there were the unshoppish experts who sometimes got a word in. V., for instance, who received a bundle of financial papers by every mail and was ready to make the fortune of anyone who would trust him with a purely nominal amount of capital as a start; Humphrey, who talked politics based on the broad general principle that all members of parliament were time-serving, hot-air merchants; Mac, who believed that pictures were entitled to be ranked as works of art only in proportion to their ability to stand the test of examination under a magnifying glass like well-focussed photographs; rather taciturn Herbert, who put in a remark here and there, but on the whole was, again in the words of Slamby, "like a fatted skeleton at the feast."

But our great stand-by was B. His subject was motor cars, and his information embraced not merely the latest types of Merc, and Rolls-Royce, but also the earliest almost prehistoric Bollée, preceded by a man carrying a red flag, which in its processional progress he had seen and noticed when, an infant, he was being wheeled out in his perambulator. He needed little encouragement to plunge into a dissertation bristling with dates and records, and he was as inexhaustible as his listeners,

ON DUTY AND OFF

with one exception, were the reverse. The exception was Humphrey, who would drop politics and join issue with a stock of personal experiences only less than B.'s because he was younger. But he could assert his facts with a fine show of authority, and to hear them together when they were both in form was to realise, so serious were they, the direct antithesis of the topical music-hall "comedy duo." The professional term for them would be, I suppose, "Back-chat tragedians."

A call to bridge in the ante-room might put an end to these discussions, but that depended a great deal on the fall of the cards. If one of the conversationalists happened to be dummy in the first hand, it gave him a clear field with the last word in the argument. After this—perhaps in justice one might say before it—we would settle down to the rigour of the game, and there were times when we sat out the island supply of electricity, which was turned off promptly at half-past ten, and finished the rubber by candle-light. I doubt whether our bridge was less or more thrilling than any other played by a mixture of beginners and older stagers who were hustled, not altogether unwillingly, into the position of experts. Willie was one of these latter, and so was V. Humphrey, Herbert and Scottie were frankly starters at a known date. W.O.W. was the best of the bunch, and he had a wonderful patience with plodding and timid partners. His comments

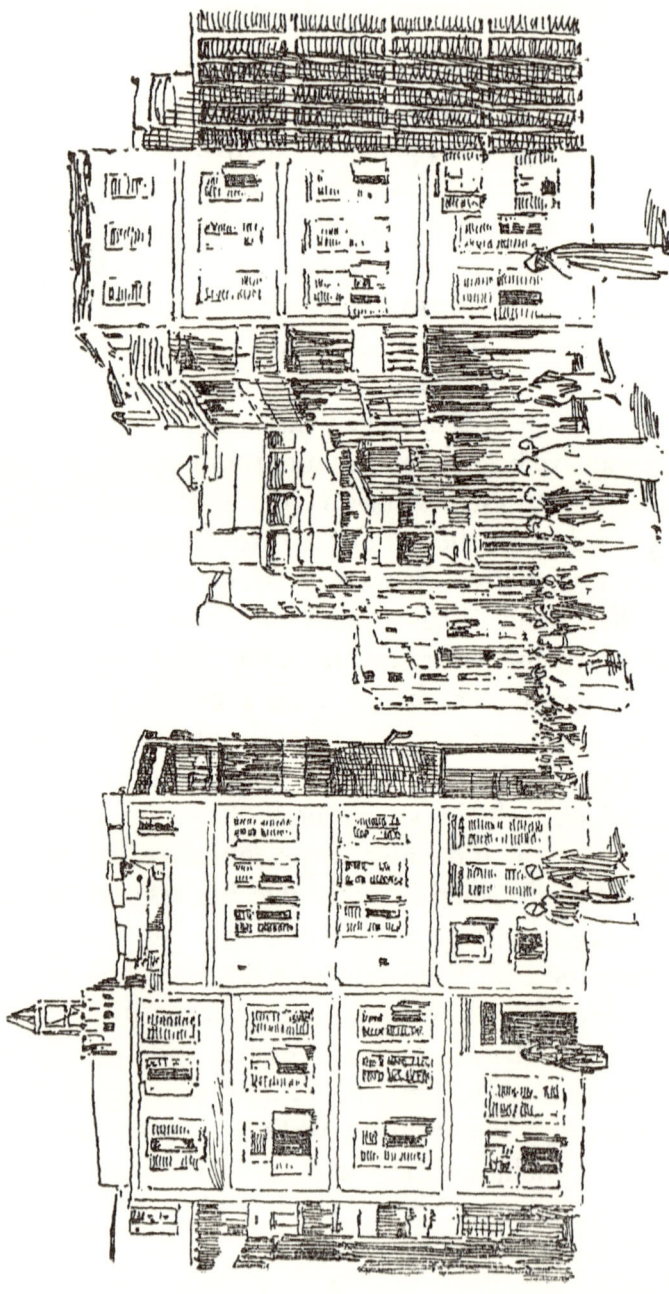

ARAB TOWN, PORT SAID.

seldom went beyond "most disasferous." Scottie played with the cautiousness of his nation, but with a delightful cheerfulness. Scotsmen either have a sense of humour highly developed or they lack it altogether. There are no half-measures. Scottie was one of the first kind. He made few jokes, but those he did make were of the best. Once having the dummy hand and finding that it contained no cards in one suit, he laid it on the table with the remark, "I have only three suits; someone must have five."

There were, of course, breaks in this sequence of events. Not everybody dined on the island every night. On Saturdays the Mess was never full, and there were other times. One of these was when Kemmy, one of the military observers, got his third pip, and we invited ourselves to dine with him ashore. Kemmy was the squadron poet. As an acrobat will throw a somersault, so he, with little or no encouragement, would throw a poem on any subject and on any occasion. But in spite of this he gave a good dinner, and after it a very cheery crowd sallied forth to wake Port Said up. One need not follow all the details of their movements. They began by capturing a hand-truck used for advertising by one of the cinematograph shows and losing it in a distant spot, and they ended by boarding a two-horse cab and driving it themselves into Arab town.

Here one of those touches of tragedy which the

East is always ready to bring to the surface pulled them up short. They came upon a crowd in the road, and the cabman, who had clung on behind his vehicle, found out what was amiss and told them in his own way: " Arab girl live at Said village. She run away to Arab man, but brother he say she come back home. She say not come. Brother he say if she not come he gibbit knife—finish. She tell one policeman and policeman he say all right no finish. Brother he come find sister, bring knife, gibbit for dead."

The policeman having arrived a minute too late was then in the act of taking the avenging brother to the Caracol. Somehow after that the " rag " lost its spirit. Perhaps it was as well that Arab town was out of bounds.

The reason for the emptying of the Mess on Saturday nights was that the Casino Palace Hotel held a dance, beginning at nine and ending at eleven or thereabouts. This weekly festival was known as the " small dance " to distinguish it from the full-sized one, which lasted frequently till the big hours of the Sunday morning. But this was not in our day. It disappeared when the war brought military law; disappeared together with the little race tables which justified the hotel's name. Many other things had gone in like manner. Indeed of Port Said before the war much the same remark might have been made that Dr. Johnson made of London: " men of curious inquiry might see in it

ON DUTY AND OFF

such modes of life as very few could ever imagine." It seems scarcely credible, for instance, that the streets were cleaned by municipal droves of pigs which were let loose at night and given the free run of the gutters. This, however, by the way. If Port Said before the war was not civilised to the pitch of employing ordinary scavengers, it at least knew about dances, and if the big ones failed to stand the strain of hostilities, the small ones were a useful substitute.

One would dine at the hotel first, in order, as Guy used to put it, to get up flying speed, and then join the revels. On the whole, I think they were as interesting to watch as to share. The music, produced by a piano and two stringed instruments, was always the same; week after week the same tunes in the same order. I don't quite know why this was. Perhaps the musicians, always the same, too, gave their best and that was their all. Perhaps the dancers preferred it so, because they could then tell exactly where they were in the programme. It may have been that, for the dancers, the ladies at least, were also always the same.

One recognised, indeed, the fair creatures who took part in the afternoon promenade past the Eastern Exchange Hotel, and we might also have seen them having tea on the Casino terrace with its view of the bathing huts and the De Lesseps statue. There were few English girls among them, and not many French, but there were many Levantines,

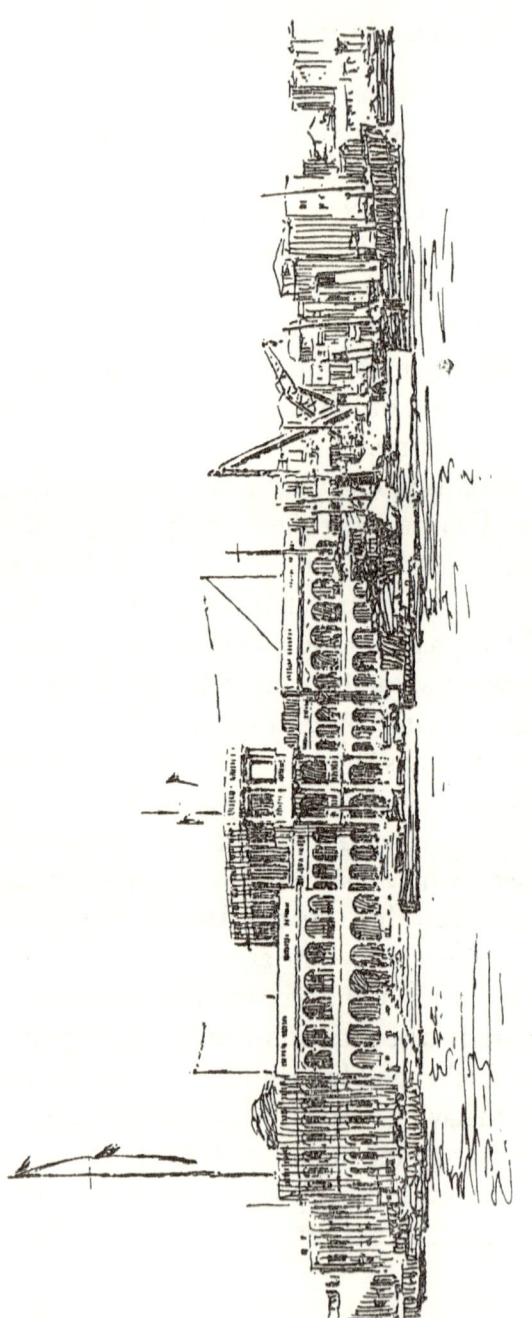

NAVY HOUSE, PORT SAID.

ON DUTY AND OFF

which might mean that their nationality was anything. Their mothers were there invariably; sometimes their fathers; generally their aunts and elder sisters; and these formed a formidable breastwork from which one's partner was extricated. She would dance vigorously, cease dancing with the music, and return to the fold. No sitting out in quiet corners; no heart-to-heart conversations; no soft nothings; no melting mood. Or, stay; I think perhaps that is hardly accurate. Anyhow, if one melted, as one felt ready to do, it was done in the family circle or prevented by the flapping of the family fans.

And yet there was a suspicion abroad that these young things were, so to say, pawns in the great game of international alliances; that military and naval officers were regarded as eligibles. It is difficult to believe, but there were one or two cautionary tales in circulation. One officer danced two dances in succession with the same lady and in the interval filled a vacant seat in the cordon of relatives. He got the idea somehow or other—based perhaps on the ordinary social amenities of Western Europe, but more particularly on the hope that it might help him to learn a foreign language—that if would be polite to call one afternoon about tea-time. He called, and found a pleasant room full of people, and was congratulating himself on having made something of a general hit, when he was aware of a phenomenon. The room became gradually and almost imperceptibly empty, and at length he found himself alone with an elderly gentleman. It was papa, and papa alone was not there to discuss the weather or the crops, or any of the lighter topics. Something weighed heavily on his chest, and he threw it off with a blunt question as to the officer's intentions. The officer thought of several witty and crushing answers, but only after he had made a hurried departure.

Another officer had an experience something similar, but he lacked a similar dexterity in retreat. How it came about he never quite knew, but it seemed to be tacitly understood that an engagement had been entered into. "A marriage," as they say in the newspapers, "has been arranged and will shortly take place"—that was about what it meant. He was allowed, nay expected, to walk in the afternoon parade, alone with the lady—alone so far as advance and flanking guards were concerned, but very strongly supported by a following escort of sisters, aunts and a mother. For a day or two it was quite a feature to be looked for from the tables in front of the Eastern Exchange Hotel, and then something happened. By a strange coincidence this officer had orders to proceed forthwith up the line. He had no time even to say good-bye. But such sad breaks did not often occur. There was nothing to break.

In deference to the requirements of the A.P.M.,

the small dance terminated at eleven. That is to say, by half-past eleven or so the last of the six stock dance tunes had been played and encored, and there was a general move homewards. It was too late for duty boats, and we from the seaplane base had to take ship and come to the island in feluccas. Native boatmen were none too keen on the trip at that hour, especially if it happened to be windy and rough, as it frequently was in the winter months. Port Said, indeed, knew all varieties of weather. I never saw snow there, though some people profess to have seen it, but I saw almost everything else in the calendar, from the damp, clammy heat of the summer days to the thick overcoat weather of the winter nights. Getting back to the island at such times was no joy-ride, and had we not lent a hand at the oars we might have spent more than one night afloat in the harbour.

The island itself, too, was not the most peaceful spot in the world when the clerk of the weather had the wind up. One could look across the canal on a day apparently fine, except for a feeling of oppression in the air and see what seemed to be a mist. Ships moored on the opposite side would fade away, and with only a few minutes' warning a campseen would be upon us driving its stinging sand into our eyes and faces, and playing havoc with our property. One of them lifted the roof clean off the Mess; another took an enormous Besonneau hangar, a heavy timber structure covered with canvas, turned it completely over and piled it in wrecked confusion on some adjoining buildings.

Such things, however, did not often happen, and our stormy passages back from the small dance served only to render more inviting the narrow packing-case cabins; and the songs which accompanied the undressing process at night, even as they did the dressing process in the morning, soon died away.

Willie, I remember, was the chief performer on these late occasions, and his one song informed us how the roses round her door made him love mother more. But one night there was no song. Returning with three or four other roisterers from the Casino, Willie entered his cabin with his usual cheery stride and encountered an obstacle. Feeling gingerly his hand grasped a chair, and he tried to withdraw it, only to find it entangled with another. The two came out and let down a third, and then Willie struck a match. By its flickering light he saw his cabin in the guise of a second-hand furniture shop. Every piece of movable furniture on the island had been piled into it. His companion revellers stood by to help him, and twenty-nine chairs—one never guessed there were so many to be found—were passed out through the window into the gangway between the hut and the Mess. Also there were tables, a huge wooden bath-tub, and many packing cases. Finally he reached his bunk, and removing the coal box, retired to rest.

ON DUTY AND OFF

We never learned with truth who was the instigator of this outrage, but we had our suspicions. Puck was the only one who, next morning, declared that he had slept undisturbed through the tumult of the removal.

VII

THE LOSS OF THE BEN-MY-CHREE

THE few operations of which I have given details must be taken as typical of the activities of the Ben-my-Chree during the busy year 1916.

She opened 1917 with an equally promising programme, but the fortunes of war were against her. Hitherto she had penetrated enemy waters and braved unscathed the fire of enemy guns and aeroplanes, but January of the new year was to see the end. Her first trip, in which she had intended to visit once more the Gulf of Alexandretta, was stopped through heavy weather which drove her for shelter into Famagusta, whence she was recalled to the Base. On 8 January she was out again with orders to undertake flights for the French at Castelorizo. The flights were never carried out.

What her exact orders were is neither here nor there, but it may be stated definitely that they were not those with which she was credited in an article published about a year later in an Australian monthly magazine. The writer of this remarkable composition stated that the Ben-my-Chree was carrying six seaplanes and the necessary staff for

THE LOSS OF THE BEN-MY-CHREE

establishing a base on the coast of Asia Minor. We may take it that there was nothing of this kind in the instructions which the Commander received from the French Admiral after the arrival of the Ben-my-Chree in Castelorizo harbour on the morning of the 9th. Part of them, at any rate, was to the effect that since weather conditions were against flying, the ship was to remain at anchor.

Whatever the remainder might have been, no one expected to have very much to do until the following day. Most of the crew were looking forward to a quiet afternoon; others not so quietly inclined were going ashore to listen to the efforts of the ship's band which had been invited to perform in the market-place. Some, in fact, had already gone ashore when, at 2.10, an explosion was heard close to the port side of the ship.

So unexpected was it that the various occupations which are understood in the phrase " make and mend " were most rudely disturbed. Sleepers were awakened; reader's put aside their books; one man who really was making and mending ran a needle into his thumb and carried it with him while he leaped to the ship's side and pushed for a place at one of the already crowded port holes.

The earliest impression was that there was an attack from the air, and it was supposed that somehow or other the Turks had produced seaplanes or land machines from somewhere, though there was no information of any being anywhere in the neighbourhood. But these conjectures were quickly dispersed by a definite realisation of the truth. The first explosion was rapidly followed by another, and the splash as the projectile struck the water close at hand made it clear that the ship was not being attacked from the air but being shelled from the mainland.

The position of the guns was immediately revealed. They were on the hills to the north-east of the island facing the entrance to the harbour, and the first thought was to hoist out seaplanes and retaliate by an aerial attack. This had been done on the last occasion, only two months before, when the Ben-my-Chree had come under fire from a land battery—it was near Adalia—and two guns out of three were hit with bombs. But nothing of that kind could be done now. The third shot came close on the heels of the second and first, and the idea had to be dismissed almost before it was entertained. For the third shot entered the hangar and set fire to it.

The Turkish gunners had got the range to a nicety and other hits followed in quick succession. The Ben-my-Chree's guns had been promptly manned, but the enemy was found to be firing at 8,000 yards with the great advantage given by the height of the hills, and the ship's guns could do nothing with any accuracy. Her steering gear was carried away; her whaler was completely destroyed; shells made her engine room untenable,

THE LOSS OF THE BEN-MY-CHREE

and all parts of her were continually hit. All this took place during a very few minutes, and three orders from the bridge following each other with only a brief interval between them, indicate the rapid progress of events. First, on the supposition of an aerial attack, came " action stations " ; then " fire stations " ; and finally " abandon ship."

With the utmost speed and coolness each of these orders was in turn obeyed, though it was not without regret that the gun crews left their guns and took their places with the fire parties. But the fire had gained too firm a hold to be got under, and quickly though it succeeded the second, the third command had already appeared as inevitable. The crew left in good order, most of them in motor boats and cutters, the remainder swimming. Here again we must correct the Australian journalist who pictures one of his compatriots treating, as a mere nothing the swim of a mile and a half " even with an aerial camera on his back." Had there been a mile and a half to swim the hero of this exploit, if we on the station were any judge of him—and we knew " Uncle Wallah " pretty well—would have abandoned the mythical service camera and lent his back to less able swimmers. There was, however, no such necessity. The shore was only about eighty yards away; the swimmers took the water more from inclination than from need, and all reached land. Forty minutes after the first shot had been fired the last boat had got away from the ship, and

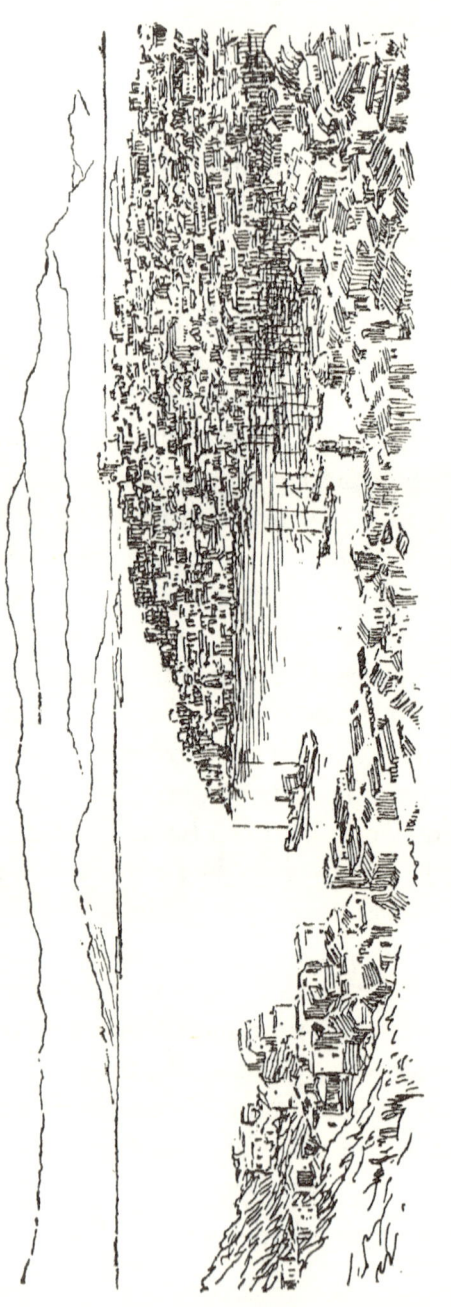

THE HARBOUR, CASTELORIZO.

THE LOSS OF THE BEN-MY-CHREE

she was left to the flames, which were now making short work of the petrol store, and to the attentions of the Turkish gunners, who continued their bombardment until the riddled hull had found bottom, and only the charred and blackened superstructure remained visible.

The shelling was intermittent till dusk—somewhere about six o'clock—and for the greater part of the four hours, the Ben-my-Chree was the target. But there were intervals during which other objectives were engaged. A French torpedo boat lying in the harbour got under way when the bombardment began and ran the gauntlet as she moved out. The French Admiralty yacht was fired at, as was a French destroyer, the Ben-my-Chree's escort. Many shells were thrown at the wireless station on the heights above the town, but the station escaped destruction. It was the Ben-my-Chree that the Turks wanted, the ringleader and the most persistent of the three harriers of her coast towns, railways, camps and lines of communication, and having found her and run her down in the snug little harbour of Castelorizo they made sure of her.

Where then and what is this Castelorizo, with its harbour and its French battleships so easily within range of Turkish guns?

" Castelorizo is so small that no one knows where it is, and no one cares." I quote these words from an anonymous booklet, entitled *L'Isle de Castelorizo* and published in Cairo in 1917. They are taken from a statement made by an emissary from the island who went to Paris in 1913 to try to induce the great European Powers to interest themselves in its condition and to save it from the Turks, who were than exercising over it a grasping and intolerable despotism. Geographically, Castelorizo is the largest of a group of small islands off the coast of Asia Minor, some 350 miles to the west of Karatash Burnu, the headland to the South of Adana. Just over four miles long and rather less than two broad, it stretches north-east and south-west less than three miles away from the south of the big drooping curve of land between Rhodes and Adalia. It possesses one town which encircles a harbour facing towards the mainland, and there is a small suburb with a second harbour. The rest of it is a series of high rocky hills, not easy of access, and its dry soil is only very sparsely cultivated.

Historically there is very much more which might be said, but it would be difficult to justify the inclusion of any lengthy account of ancient and mediæval struggles among records of twentieth century warfare. Let me hope that a very brief survey will save me from the stigma of disrespect for storied ground. A Greek inscription dating from the third or fourth century B.C. attests the existence of a very much older castle than that of which the ruins still look down upon the roadstead. Roman historians speak of the harbour as spacious enough to contain a fleet. The Knights

THE LOSS OF THE BEN-MY-CHREE

of St. John of Jerusalem had possession soon after 1300, and built a fortress, which in 1440 was attacked and laid in ruins by the Sultan of Egypt, who claimed the islands in this quarter as part of his empire. In 1450 the Pope, in the interests of the Christian inhabitants, ordered the rebuilding of the castle, and the island a few years later seems to have been ceded by the Knights of St. John to the King of Naples, to whom the Pontiff had entrusted the work, which was completed in 1461. The Turks sacked it as a side issue of an expedition against Rhodes in 1480, but their ruler, Mahomet II., died before his ambitions were realised, and the island remained in the hands of the King of Naples. About 1630 the Turks got possession of it, but the Venetians wrested it from them in 1659, and the castle was once more destroyed. After this the inhabitants appear to have been left in peace for upwards of 150 years. They were for the most part sea-faring. They traded as far as Alexandria, but lived largely on the products of their fishing, and early in the nineteenth century this lack of external interference was apparently beginning to have an ill effect. The island had earned a reputation as an independent nest of freebooters.

Lieutenant-Commander D. E. Hogarth, in a note in *L'Isle de Castelorizo*, from which I have condensed the above very sketchy account, states that there is in existence a report by the captain of a British vessel which was captured about 1820 by the pirates of Castelorizo and taken to the island, where captain and crew were brought before a General Assembly and eventually released with their ship. In 1833 the island was ceded with others by Greece to Turkey, and then began a series of oppressive enactments which culminated in 1898 in an appeal to the Powers for intervention. This was followed in 1913 by the sending of the delegate already mentioned to Paris. But the Powers were powerless, and Castelorizo remained under the Turkish yoke, though a Greek Governor was still in control at the outbreak of war in 1914, and when food difficulties brought about a revolution it was a Greek cruiser that set out thither with a new governor and a party of soldiers. The French naval authorities, whose duty it was to patrol the coast of the eastern Mediterranean, intervened, however. Two French cruisers anchored in the port, landed a force which occupied the island, and hoisted the tricolour over the crumbling battlements of the castle.

It was, then, to the town of Castelorizo—town and island bear the same name—that the ship's company of the Ben-my-Chree made their way during the first hour of the bombardment. They found the steep, narrow, tortuous streets thronged with inhabitants in a state approaching panic, which the French Governor and his marines were for the moment almost powerless to restrain. Such a state of affairs was not unreasonable. Hitherto the war had shown this people none of its horrors.

THE LOSS OF THE BEN-MY-CHREE

On the contrary, with the advent of our Allies had come a vast amelioration of their conditions of life. Their food supply was assured; there was relief from the Turkish burden of ruthless and indiscriminate taxation; public health was giving evidence of the control of a highly skilled medical officer, who already had in check the frequent scourge of epidemic disease.

But now, suddenly and without warning, the war was at their very doors. The Turks had laid their guns well, but many shells flew wide of the mark. In particular the houses suffered which fringe the harbour near the spot where the Ben-my-Chree was at anchor, and the little square houses of Castelorizo with their rough courses of stone, poorly mortared and with horizontal layers of timber none too sound, are not built to withstand much knocking about. And, as we have seen, the Ben-my-Chree was not the only objective, though she was the chief one. Many of the shells fired at the wireless station to the east of the harbour made havoc in civilian dwellings, and of those aimed to arrest the progress of the other ships as they made their way out of range, many more grazed the rooftops of the houses straggling down the hillside to the water's edge.

But there is a curious feature of Castelorizo which to the eyes of the strangers who came ashore made it appear that the bombardment had done far more damage than was really the case.

It is a town which is continually growing, and its growth is due to a quaint custom. Whenever a girl is born, her father immediately lays the foundation of a house which, in due course, shall be hers to live in. At intervals during her life, if she lives, work is added until finally on her marriage day the residence is ready. But though many houses are begun, many never rise above the first few fragments of stone, and others are left in various stages of incompleteness. Thus, dotted about among the inhabited dwellings, one sees ruinous structures which, to the unenlightened newcomers, looked like the work of the missiles which had been the cause of their coming.

There was, however, no leisure for sightseeing, even had anyone at this moment been much inclined for it. There was work to be done with the boats and in conveying to safe places such gear as it had been possible to put aboard them. There was not much of this, but it included several of those articles which in an abandoned ship are the first thought of responsible persons. The officers and liberty men who had already gone ashore before the disaster were on the quays at first to watch—they could do no more—and afterwards to help. They were in the heart of the town when they had heard the first explosions, and, hastily summoned by a French look-out from the wireless station who told them that their ship was on fire, they had run break-neck down the winding

THE LOSS OF THE BEN-MY-CHREE

alley-ways. Arrived in view of the Ben-my-Chree, they had been in time, standing in the lee of the hill over which the shells were flying, to see the departure of the French destroyer as she slowly left her moorings, slowly got up steam, slowly drew out of reach of the guns which had churned up many vicious splashes in her path. This dramatic passing took place between them and their own ship, from which the dense black smoke of burning petrol formed a haze over the rocky slopes opposite. But this sight, too, there was little time to watch. The ship's boats were busy and several journeys had to be made. At length, however, all that could be done had been done. There was a muster, and orders were issued for all to assemble at four o'clock in the market-place where, if things had not been as they were, the ship's band should have been giving the local audience an insight into R.N.A.S. ragtime.

There was no music in the picturesque old market-place, but the ragtime element was not entirely lacking from that four o'clock gathering. Through the crowd of quaintly costumed Greeks, by no means yet reassured, though the French sailors were doing their best to cheer them up, despite the continued menace of bursting shells, came the ship's company of the Ben-my-Chree in all conceivable varieties of clothing. There were the working garments which may be seen in any ship on active service, but in which no self-respecting sailorman allows himself to be seen ashore, and between these and the regulation clothing of those who had landed early in the afternoon there were many shades of differentiation. There was the shore-going rig of those who had had time to change and whose pockets had the bulging look of those who believe that it is better to save too much than too little. There was the half-and-half or quarter and three-quarters get-up of those who had been able only to snatch an ill-considered supplement to the garments in which they were working. There were those of the swimmers who, in the excitement of the moment, had flung aside encumbrances which they would have been glad enough now to add to the dank and clinging residue. And there was that of the Flight Commander who had left the ship last of all. The final boat-load had forgotten him down below making a leisurely inspection to assure himself that he really was the last. He had swum ashore and landed near a native cottage. The inhabitants had rushed to his assistance and in an excess of hospitality had divested him of his dripping uniform and supplied him with a costume of local design. His knowledge of modern Greek was limited to two words which he couldn't pronounce and they couldn't understand, and no entreaties could dissuade them from their purpose of drying his own clothes and decking him out in theirs. It was in this rig that he paraded in the market-place.

THE LOSS OF THE BEN-MY-CHREE

But if the setting and costumes were fit for a musical comedy or even for a comic opera, there was no such light note about the business in hand. Many of the townsfolk whose curiosity had triumphed over their fears thronged the marketplace and the narrow streets leading into it, or peeped from the windows of their houses. Many more, waiting the stimulus of active leadership, were wandering distractedly in the by-ways, clinging to each other in terror at each fresh explosion, gazing fascinated at the strip of sky visible between upper storeys in which opposite neighbours could clasp hands. Others, again, taking their children and a few belongings, had fled to the surrounding heights where, crouching in the crevices of the rock-faced hillsides, they watched and waited for they knew not what. It was with this condition of affairs not less than with the accommodation of the officers and crew of the burning ship that the Governor of Castelorizo had to deal, and he handled the whole situation with a tact and cool-headedness which gradually restored order and comparative peace of mind.

The intentions of the Turks were not known, and surmise had to take into consideration the worst imaginable developments. It was clear that they intended to sink the Ben-my-Chree, but what their further intentions might be could not be so definitely determined. They were attacking the wireless telegraph station, and in the process were doing indiscriminate damage to civilian property.

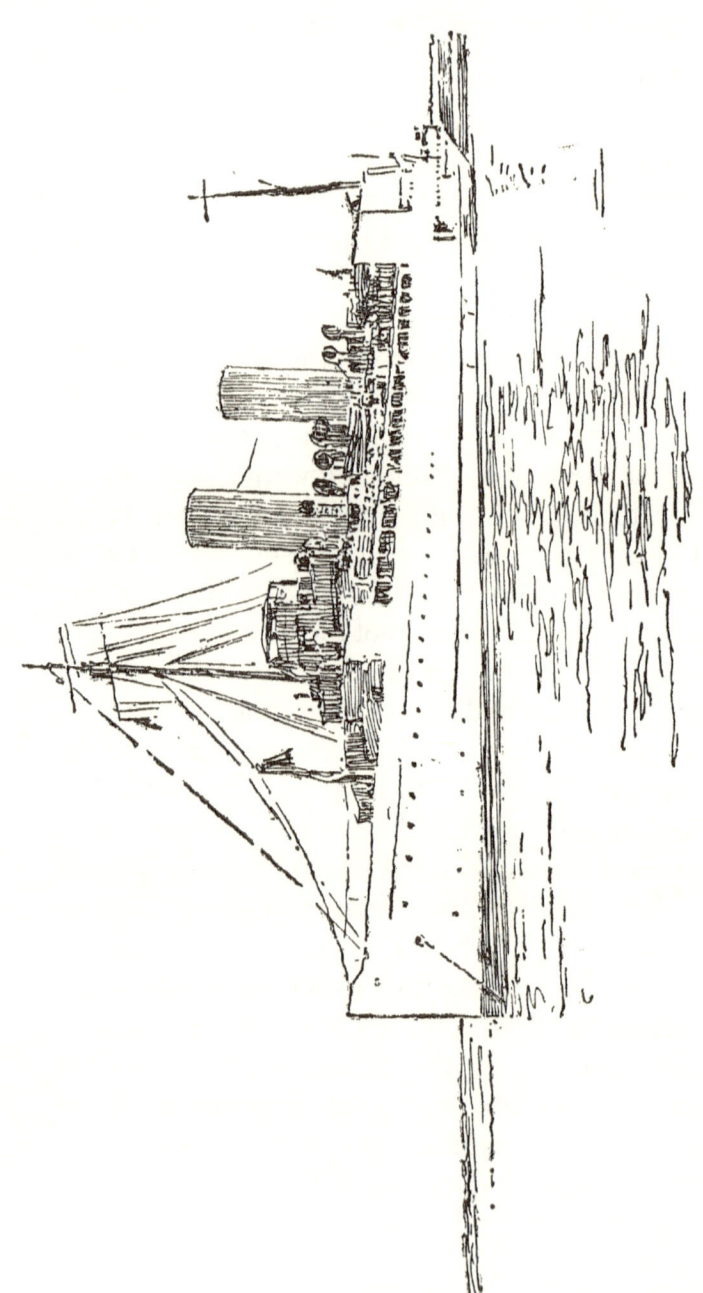

H.M.S. BEN-MY-CHREE.

THE LOSS OF THE BEN-MY-CHREE

Was this the prelude to a systematic bombardment of the town? Was it the enemy's object to retake the island, which had been a thorn in his side since the day of its occupation by the French? This at least was within the bounds of possibility, even of probability; and it was with this in view that the Governor framed his plan. The inhabitants were ordered to leave the town and to make, not, as the earlier fugitives had done, for any spot in the surrounding country which appeared to offer refuge, but for definite points where their rationing could be properly supervised. The instructions were conveyed and put into effect as quickly as might be, and some semblance of order was restored, but it was far into the night before the evacuation was complete. Not until the shelling had ceased with the dusk did many dwellers find the courage to exchange the precarious shelter of the town for the safety of the hillside out of range of the guns.

Meanwhile the ship's company had received instructions, too. The Governor had pressed them one and all into service as auxiliaries to his small force of French sailors to whom would fall the defence of the island in case of attack. They were divided into parties, and allotted posts to which guides directed them. Three of these posts were above the town and near at hand. Here they arranged watches and kept an all-night look-out. To a fourth party were assigned duties which gave a prospect of more excitement than that promised by sentry work.

ABOVE AND BEYOND PALESTINE

This fourth party joined with a contingent of French sailors and took up a position on the coast facing the Turkish mainland to oppose any attempt at landing. They stayed there several days, but no landing was actually attempted, and I think the Franco-British defenders were rather disappointed than relieved. W., the senior officer of the military observers, was in command of the detachment, and he introduced several ideas which were responsible for a feeling of genuine regret when the Ben-my-Chree men at length parted from their allies to make the voyage home. He was the only military officer there—the only officer at all, in fact—and his servant was the only soldier among the " other ranks," and together, since the post was there for military purposes rather than naval, they organised the ordinary military routine of trench digging, drills, guards and fatigues.

I say " they " advisedly. Private Bedford, as I think he would prefer me to miscall him, was a peculiar type of the new British soldier. He was capable and efficient, but he came of a civilian stock which held the view that social inequality started from itself downwards. This meant that Private Bedford considered himself as good as the next man, whoever he might be, and probably better. In a quite friendly way, mind; there was no truculent self-assertiveness about him. The disciplinary conception of rank made no appeal to him, and that is possibly why he was never an N.C.O. The word " Sir "

THE LOSS OF THE BEN-MY-CHREE

did not figure in his vocabulary. He addressed and spoke to officers as " Mr. So-and-so," whether they were subalterns, captains, majors or colonels. It is not on record that he had any dealings with generals, but doubtless he would have made them no exception to his rule. It is related of him that on being told off once at the seaplane Base on some minor charge and asked if he had anything to say, he replied, " There ain't many of us 'ere. What I say is let's live together in 'armony." He had the British better-class mechanic's pride in doing work well, and the orders given him as an officer's servant were generally carried out. If he failed he was liable to provide an excuse which disarmed criticism. Once, having been instructed to wake W. at 6, he woke him at 7.45 with the pleasant remark, " There, I've let you lay a nice long time this morning, 'aven't I ? " One of his assets was an imperturbable cheeriness which found frequent expression in a low windy whistle which occasionally, but not often, suggested some popular air. It was this cheeriness, coupled with the discovery that he was far more skilled than any one else at rifle drill, that gave him in that little corner of Castelorizo the pre-eminence which I have indicated by coupling him with his commanding officer. Another of his assets was the ability to get other people helping by sheer force of good fellowship. W. was awakened at the Base one morning at the right time, but not by Bedford. This being pointed out to him, he explained " Yes,

I got 'im to come along. But "—with an engaging smile—" I was close on 'is 'eels."

This kaleidoscopic mixture of positive and negative qualities does not perhaps suggest the ideal helper in a task which required immediate organisation and had to be begun in gathering darkness on ground which was strange to many of those engaged in it, but the fact is that Private Bedford certainly took a prominent and capable part, and everybody else entered into the adventure with such zeal that it developed, in spite of the underlying seriousness of purpose and not a few bodily discomforts, into something resembling an extraordinarily well-managed picnic in which every one was doing exactly what pleased him most.

Two of W.'s ideas in particular assisted largely towards this condition of affairs. He spoke French fluently, so that so far as he himself was concerned the bi-lingual character of his command presented no difficulties. He could give his orders in the two languages and see them carried out, but this meant, for all the mutual good fellowship that existed, a division into two camps, and he saw that the removal of the barrier would make for efficiency. At a parade, therefore, each morning and each evening, Frenchmen and Englishmen were drawn up in two lines facing each other, and every man in turn stepped forward two paces and called out his own name aloud. It was not long before everybody knew everybody else. This was the first idea. The second was the institution of French lessons. They were not compulsory, but every man attended them, and their popularity was readily observable afterwards when the pupils could be heard testing on each other the extent of their acquirements. Even Private Bedford, who, like Jane, the parlour-maid in one of Mrs. George Wemyss' delightful works, was rather proud of speaking a language which many of his companions didn't understand, sank his insular prejudices and expressed his approval in the phrase, "*Oui, très bon.*"

It was not until 21 January that this party, taking passage in a friendly trawler, arrived back in Port Said, but the others got away earlier. Some of them stayed on the island of Castelorizo three or four days; others were able to embark on the day following the disaster. The bombardment had ceased, but no one knew whether at any moment it might not be resumed, and it was obviously unsafe to use the harbour, under cover of night, therefore, the ship which carried them lay off the craggy coast at the far side from the mainland, and thither the officers and men, together with four of the crew who had been wounded by fragments of shell, made their cautious way. In perfect silence a chain of men was formed down the steep sipping face of the rock, and from hand to hand the wounded were passed in safety to the waiting boats. It was by no means the easiest of many unaccustomed tasks which the ship's company had been called upon to perform.

VIII

SOMEWHERE EAST OF SUEZ

At the Base definite news of the loss of the Ben-my-Chree was preceded by many circumstantial rumours. At first not much account was taken of them. They formed a sequel to equally circumstantial rumours which spoke of a deep-laid plot for her destruction. How such rumours came into being I know not, nor was it worth while to inquire. They were always there, and they appeared as a lurid comment on the workings of the native mind. Old stagers at the station paid little heed to them, and newcomers soon grew out of the mood of puzzled credulity. There were, it is true, people who were prepared to convince any one ready to be convinced that their origin could be traced to some mysterious agency akin to those which form the subject of uncanny tales of the Central African tribes. But I think the Port Said news service, if the gossip-mongers were to be believed, was well ahead of that of Central Africa, whereas the Kaffirs were reputed to know of events in an inconceivably short space of time after they had actually taken place, we in Port Said knew of them simultaneously with their happening. Indeed, not infrequently the news arrived before the event had a chance of taking place; and since as a rule the events referred to never did take place at all, the news service had everything its own way.

In a sense it was something like a three-act play, with the last act missing. The first act would foretell misfortune, the second would describe it, and the third ought to have explained away all difficulties to every one's satisfaction. Sometimes the play did not get beyond the first act, but that was invariably presented. I can remember hardly a single occasion of the departure of a seaplane carrier on some coastal expedition without a prelude of persistent whisperings foreboding disaster. Some one, perhaps, had been having his boots cleaned by one of the Arab boys who plied their trade with a home-made-looking outfit in front of the Eastern Exchange Hotel. He would be seated in the row of little tables sipping coffee while the ragged youngster scrubbed at his feet, when he would be approached by a native hawker of cigarettes or picture post-cards or coral beads or walking-sticks of "rhinoceros" hide (which was really hippopotamus), under cover of his demonstrative and persuasive salesmanship, encouraged possibly by a successful deal, the hawker would convey the fate-laden words: "Raven ship go out. No come back. Mafeesh."

Always, as I say, there was some such prophecy

as this for the first act. Sometimes the second appeared in statements purporting to prove that the prophecy had been fulfilled. And then the ship would come gaily back to port and no one worried about the third act being missing.

The departure of the Ben-my-Chree on her last voyage was no exception to the rule. There were the usual fictional accompaniments and the usual chaffing reception of them, but the pleasantries gradually died away when a day or two after her setting out, the curtain rose on a particularly realistic second act. The story of the ship's fate crept gradually from the region of romance into that of reality, and soon, though there were many conjectures as to what the third act might bring forth, there was no longer any question that there would be one. That hints of the tragedy did actually arrive at Port Said within a very few hours of its enactment is easily accounted for without any help from the occult and incredible. As we have seen, there was a wireless station at Castelorizo which the Turks did not destroy, though they did their best. This fact possibly did not occur to those at the Base. Possibly they did not know of it; possibly many of them did not even know that the Ben-my-Chree had gone to Castelorizo. At any rate, the hints provided an unusually sound basis for a second act bristling with complications of battle, murder and sudden death. Rumour was given a free run with every encouragement which could be suggested by imaginations stimulated by spy literature marked at 7*d*. to 1*s*. a volume and purchasable in Port Said at anything up to 3*s*.

The curtain rang down on the second act with the unmistakable news that the Ben-my-Chree was lost. It was official and it gave very few details, so that although there was much left for the third act to tell, it was dear that it offered no prospect of a happy ending. It took several days in the playing and the curtain fell finally on the dramatic and authentic accounts of the first batch of survivors.

Having lost the accommodation of the ship's cabins, they had to find temporary quarters ashore, and one came upon them unexpectedly in the town refreshing themselves at the expense of eager circles of listeners to whom they were apologising for the curious cut of their hastily purchased ready-made clothing. This opening to the story seemed invariable. The sinking of the ship had already faded into insignificance in their minds in the shadow of the big fact that they possessed only the gear in which they had paraded in the market place of Castelorizo on the eventful afternoon. To some of them it was not the least trying of their experiences that they should have been compelled to face the world of Port Said in reach-me-downs. Luckily one of them still retained his eye-glass. An eye-glass gives an air of great distinction to a party whose uniforms are full of unexpected bulges and wrongly-placed creases.

ABOVE AND BEYOND PALESTINE

If I were not tired of the analogy of the three-act play, it might be suggested that there was an epilogue, continuing the action in a detached sort of way which had a tendency to alienate the hitherto rapt attention of the spectators. Perhaps it will be better to describe it as a rather tame sequel. The ship's company of the Ben-my-Chree, deprived of their working-place, invaded the island and turned-to where they could. Their paymaster swooped down upon any spare chairs and tables, and, having seized those, upon anything which might be endowed with the temporary semblance and use of chairs and tables, and collecting his spoil in a hut which seemed to him to be tinder-crowded, set his staff of writers to the task of straightening up accounts. Other people did other things doubtless equally valuable, but there was generally a sense of listlessness traceable to the prevalent idea that shipwrecked naval officers and ratings are entitled to an interim dividend of home leave. A good many of them got it, or at least they went home pending their appointment to a new ship. Others, not being actually borne on the books of the Ben-my-Chree, though they were serving in her, merely became merged once more in the personnel of the station. Others, again, contrived by various processes of submissions and representations to avoid the change which might well provide them with far less interesting work than they had then in prospect. (As I have hinted, there were many dull things to be done at home by seaplanes.) The Squadron, in short, carried on, but for some little time there was a feeling akin to convalescence after a severe amputation.

Very suddenly, at the beginning of March, this feeling was dissipated. Operation orders arrived, and the station was instantly galvanised into action. So far as concerned the rank and file and many of the officers, the orders were as vague as they had ever been. They assumed a character of unusual urgency from the fact that they followed a period of several weeks, previously unprecedented, of comparative stagnation—several weeks of fierce engine-shop energy, trueing-up, test-flights, and routine generally, but of nothing doing at sea—but they possessed a character of unmistakably real urgency as well because they demanded very big things in the way of preparation and gave a very short time for their accomplishment.

I think from first to last three days covered it. During that time material in stupendous quantities was loaded in the Raven. Machines, erected and still in their packing cases ; spare engines, spare propellers, spare floats, spares of all sorts, made the passage from the island to the ship in a constant succession of tugs and lighters. The Stores and Engineer Officers had the time of their lives, and recalled in their brief moments of relief the distant days of mobilisation.

And there was a further feature which stamped

ARABS HAULING IN A SEAPLANE, PORT SAID.

the affair as unique. One of the officers was entrusted with the task of providing a supplementary larder for the mess. This was distinctly unparalleled. For short voyages up the Syrian coast, and even for longer ones along the shores of Arabia, it had been usual to rely on the ship's steward. Evidently there was a chance now of the ship's steward making an under-estimate. The work of prevention of this possible catastrophe was done no less promptly than the other things which required doing in spite of the numerous gastronomic suggestions of other members of the party told off for the expedition.

More intimate personal preparations were carried forward simultaneously. Neither officers nor men detailed for the trip were given particulars as to their destination (or they were given them in strict confidence), but the conviction gradually grew in intensity among those not detailed that they were bound for tropical waters in the Red Sea direction. There were sighs which could be read by those versed in such matters. For trips up the Syrian coast officers' servants seldom had orders to pack camp bedsteads; for trips down the Red Sea always. Frequently a camp bed on deck offered the only chance, and that not of the best, of sleep at night. This was one of the signs. Another was the anxiety displayed in many quarters to exchange the cosmopolitan coinage of Port Said, which accepted almost any recognisable European money, for some currency more likely to be negotiable farther east. There was talk of rupees and there were hints of Mesopotamia.

It was in the morning of 10 March, 1917, that the Raven left her berth, and the suspicions of her eastern objective were confirmed when she was seen to make her way down the canal in the direction of Suez. She left behind on the island a small, rather low-spirited party, of whom the more pessimistic declared their firm belief that this was the beginning of the end of the seaplane Base at Port Said. It was perfectly clear, said these wiseacres, that the whole outfit was being broken up. What remained of the station was to be packed off home, and the Raven was going to establish a base at some spot whose position, according to the imagination of the speaker who held the floor at the moment, might be anywhere between the Gulf of Akaba and China. In any case, it was perfectly certain that there would never be another seaplane carrier stunt in the eastern Mediterranean. Though this was uttered in the face of the fact that the Anne was lying at her moorings as ready as ever for anything that might come along, it carried some weight; but very shortly after the Raven's start it got the lie direct in orders which resulted in a reconnaissance, on 21 March, over Haifa, El Fule, Nazareth and Baisan.

After this the island gradually recovered from its fit of depression, and the inhabitants forgot to think of themselves as derelicts. The routine was resumed on a fresh basis which allowed for the gaps left by the departed, and it was discovered that there were only three bridge players. I regard it as a high tribute to the fortitude with which this trying period was faced, that a new military observer was immediately put into training, and that the three allowed him to make a fourth (for extra low stakes, since one of them had to be his partner) on the second night of his novitiate.

So much for the Base. And what about the Raven? The historian, if the writer of so rambling a chronicle as this may arrogate to himself so exalted a title, may claim the privilege of being, like the bird of the anecdote, in two places at once, and he is able to lift the veil of mystery which covered her movements. Or, if not to lift it, at any rate to raise a corner of it. The Raven was on a mission which took her to a great many places, some already familiar, others very much the reverse. Her ultimate destination was Colombo, where a small temporary station was formed, but the journey thither was not hurried, and advantage was taken of it to make extended reconnaissances over the Maldive and Laccadive Islands, which form a long coral reef to the west of the coasts of India and Ceylon.

These reconnaissances were, of course, friendly. Such of these islands as are inhabited—the very large majority of them are not—were under native rulers, whose feelings were essentially with

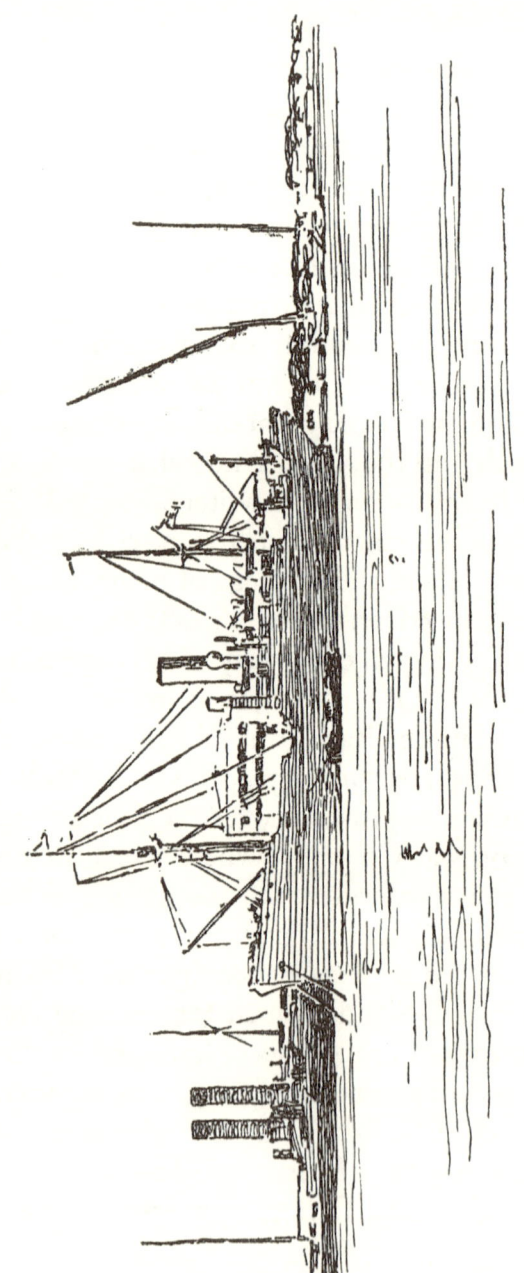

H.M.S. RAVEN II.

the Allies. But among the unoccupied islets it might well be that one might encounter signs of enemy machinations. From time to time there were rumours of hostile ships in the waters which lie between Africa and Australia, and one could never tell but that some of these desolate spots might have been turned to their ends. One might even encounter an actual raider. At any rate, it was worth while to make use of this long journey of the seaplane carrier to get into touch with native potentates and to effect possible improvements in their means of reporting suspicious intruders.

It came about, then, that the Raven, having put in a day or two at Aden on the old game of observing and bomb dropping over the Turkish positions in that neighbourhood, proceeded, in company with the French cruiser Pothuau, on a voyage of discovery which was not restricted to inspection from seaplanes overhead. Several of the islands were visited by shore-going parties, and their experiences were duly recorded. One visit may be taken as typical of several to the smaller islands.

Chetlat Island, in the Laccadives, was found to have a population of rather less than 1,000, chiefly engaged in fishing. They were not in any way interested in money, but readily accepted gifts of tobacco. The women, though unveiled, exhibited no feminine curiosity in regard to the strangers, and only two or three of them were seen. The houses were tolerably well-built structures of coconut wood with thatched roofs made from palm leaves. There was an imposing mosque with a large ornamental lake in front of it, surrounded by steps. The headman, who visited the Raven, spoke Hindustani. Poultry and goats provided most of the food apart from dried fish, and there was an ample supply of good water from deep, excellently made wells. Coconuts were the staple produce.

On 21 April a machine set out from the ship soon after four in the afternoon to make a reconnaissance in the area of Ari Atoll. At a quarter to six this machine had not returned, and the seaplane carrier and the French cruiser set about making a search in the direction of the flight. Throughout the night the search was continued without success. Hitherto during the voyage the ships had shown no lights, but now the cruiser's searchlight swept the sea with its ray. Officers took special watches in rotation, and once a distant Verrey's light was reported. It was doubtless due to imagination—imagination in such periods of high tension plays strange tricks—for investigation by boat revealed nothing. In the morning another seaplane was hoisted out, and from this a large patch of oil was sighted, and something was observed to be floating on the water. The Raven was signalled and approached the spot, and a boat was lowered. The floating object was found to be a ship's biscuit, and as biscuit was always carried in the seaplanes, it was supposed that the missing machine must have sunk at this spot, to which the strong current might have drifted it, supposing, as was conjectured, it had been compelled to "land" with engine trouble. The search for the missing seaplane was continued till the afternoon, and the ship, with the reluctant feeling that the quest was hopeless, anchored for the night off Male Island, the residence of the Sultan of the group.

There can be few things more distressing than the sensation of impotence and baffled effort experienced by those on board the parent ship in the presence of a disaster of this kind. The seaplane had been last seen fifteen miles away, apparently making her prearranged course. Thereafter there was absolutely nothing to guide a search. Had she not been driven from her course she must have been found somewhere on it. Having been driven from it by some unknown and incalculable agency of wind or weather or of engine caprice, it was impossible to tell in which direction she might be. One could only formulate probabilities and search the most likely sections of a huge area containing many possibilities of success which could not be tested. All that could be done was done, and it failed. There was nothing left but to continue the voyage, with constant inquiries at the islands on the way.

At Male the Sultan sent his secretary on board the Raven and he, speaking good English, surprised the ship's company with his knowledge of the progress of the war, gathered from British newspapers, and by his particularly intimate acquaintance with that portion of the war in which the ship was then engaged. He knew exactly the route she had taken in the north Maldives. Reports from the headmen of the islands visited had already been duly forwarded, and it was found that in some cases there was a suspicion that the Raven was an enemy ship (as, in fact, she used to be) carrying Germans in disguise. He was informed of the lost seaplane and undertook that the missing officers, if found, should be brought immediately to Male.

The following day an official visit was paid to the Sultan and a tour of the island was made. The population of Male was given as 5,000, and that of the whole Maldive Archipelago as 72,000, and it was estimated that the majority of the 5,000 contrived to have a peep at their Sultan's white guests. Curiosity was by no means confined to the men. The women stared with the best of them, and it was possible to get a quite circumstantial inventory of their personal adornments. They were all dressed more or less according to pattern. They had an upper garment of check in red and black or blue and black reaching almost to the knees, and a skirt of black cotton material with horizontal stripes towards its lower border. The hair, brushed away from the forehead, and done up at the back or side of the head, was tightly confined in a dark-coloured bag not unlike a small bathing cap. There was a good show of gold and silver bracelets, but no earrings or nose ornaments. Several pairs of nautical eyes observed these details closely, and a description compiled from their varied recollections was duly recorded in the official report. There was also this candid comment: " In general their features were plain and they were distinctly unattractive in appearance." This in the light of future events will be found to have a peculiar significance.

The town gave everywhere an impression of cleanliness and good order. Though it had evidently grown from a small nucleus dating from fairly distant times and showed the irregularity of plan which is usual in such cases, there appeared to have been later efforts to correct this with long straight roads of white coral sand, intersecting the heart and leading into the outskirts. These roads were provided with street lamps of European pattern, and the lamp-posts—a very civilised touch—were fitted with baskets of wire for rubbish. Here and there, too, were grass open spaces with groups of flowering trees. Behind these main streets there were lanes of a frequently puzzling intricacy winding between fenced-in gardens and thatched dwellings. A few of the larger houses were of coral, stone and wood, but there were many more of wood alone, with interwoven palm leaves for the roofs. These had occasionally a galvanised iron protection against fire. Every garden had a deck chair or a swing, or both, and every garden fence had near its entrance a knocker, made in most cases of any suitable piece of scrap iron. There were no vehicles, for no distance was too great for walking, and commerce did not deal in bulk. What one bought one could carry.

The Sultan's palace was a stone building, comparatively substantial, but with few architectural pretensions. Of such, however, a few traces were to be found in the religious edifices, of which several were of respectable antiquity. In particular the principal mosque, a smaller one built by Buddhists, and the tomb of Tabriz, the Persian who brought the Mohammedan faith to the island, showed some quite interesting specimens of stone carving.

The few shops in the island were owned for the most part by native merchants from India and Ceylon. Some of these tradesmen spoke English, though their wares were scarcely calculated to attract a large English custom. They displayed such articles as dried fish, betel nut, areca nut, chilies, sweets in brightly coloured paper, cheap crockery and earthenware pots, cotton goods and mats, and ironmongery in the form of nails, screws and knives. There was also kerosene for lighting purposes.

The staple diet appeared to be fish and rice; but to inhabitants of epicurean tastes a more varied menu was available. There were cattle, small in size and perhaps not the worse eating on that account, fowls and ducks, a few sheep and many goats. Coconuts, of course, were plentiful. The pumpkin was the chief vegetable grown, and there was some cultivation of bananas, bread fruit and limes.

The making of mats and coir ropes had at one time been the leading industry, but this had died out. The Raven party found a quite considerable activity in the building of surf boats of all sizes. Dried fish and coconuts were the principal articles of commerce, and for both the best market was found in Calcutta.

Before the departure of the seaplane carrier the Sultan sent as gifts a young bull, two goats, two turtles, many coconuts, bananas and limes, and a plentiful supply of fresh fish and eggs—all very welcome. And each officer was presented with a mat of native workmanship made of fibre very finely woven and of designs strongly influenced by the proximity of India. Thus laden with tokens of high favour, the Raven continued her course to Colombo, where she arrived on 25 April.

On 6 May a boat towed alongside, and two bearded and curiously dressed figures came up the accommodation ladder. They wore tarbooshes, white duck trousers and scarlet jackets with white facings. They looked like the leaders of the local band going round for subscriptions, but they were not. They were Guy the pilot and Willie the observer of the missing seaplane!

I know of no precedent for a historian (with

apologies as above) being in three places at once, and the adventures of the two officers must be taken on their own responsibility, since they were the only eye-witnesses of their exploits. Nor is it necessary that I should attempt to reduce to order the hilarious and incoherent versions which both of them gave simultaneously within a few minutes of their return to the ship. They continued to give them on demand—one could not hear them too often—for several days afterwards, and from these subsequent narratives a tolerably accurate record might be compiled. But even this labour is spared me. After the first excitement had died down each of them wrote out an account. Guy sent his home in a letter, and Willie stored his up for future reference. In the course of time each version achieved the publicity of a popular magazine, Willie's under his own name and the title "Our Seaplane Adventure," and Guy's embodied in a delightful yarn by Mr. Rudyard Kipling. Mr. Kipling, who must have got hold of the letter somehow—these writing men crop up everywhere—called his essay, what, indeed it was, "A Flight of Fact."

Willie wrote, as a good observer should, a clear, plain statement of what occurred, admitting no embellishments other than one or two lyrical touches, excusable in the circumstances, about "foam-tipped waves" during the swim ashore, and "crashes of thunder" and "vivid flashes of violet flame that cut the heavens asunder" when the lightning came on. Mr. Kipling, giving the same essential information (with one addition, which we'll note later), lavished upon it the wealth of his magic, and critics of literary technique may marvel at the ingenuity with which he shifts responsibility for the improbable and yet endows the story with all the plausibility of good fiction, which to be really interesting must be more plausible than mere fact. He introduces Guy's letter into the story in a conversation between three naval officers in a destroyer, Jerry Marlett, Augustus Holwell Rayne, *alias* " The Damper," and H. R. Duckett. Marlett mentions a visit to All the Pelungas, and Duckett says his acting sub-lieutenant has a cousin who had been flying there—" a Loot called Baxter." Duckett tells the story from his recollection of his acting sub's recollection of a letter from Baxter, so there is plenty of scope for excuses if any detail appears to be unlikely or inaccurate. The narrator is closely cross-examined, and when he cannot explain he can answer " I'm only tellin' you what my sub told me. Baxter wrote it all home to his people, and the letters have been passed round the family."

On the question, for instance, of why on earth anybody should be flying at all in such a region:

" 'As far as I can make out,' said Duckett, 'Lootenant Baxter was flyin' in those parts—with an observer—out of a ship.'

" 'Yes, but what *for*?' Jerry insisted. 'And what ship?'

ABOVE AND BEYOND PALESTINE

" 'He was flyin' for exercise, I suppose, an' his ship was the Cormorang. D'you feel wiser?' "

I should steal Mr. Kipling's story bodily and insert it here were it not that my purse contains nothing of greater value than a conscience which warns me against committing breaches of copyright.

Here, however, are the dry bones of the narrative gathered from all possible sources, conversational and documentary:

The machine had been blown out of her course by a sudden wind-storm; rain came on, darkness was approaching and the supply of petrol was getting low, so a landing had to be made near one of the islands, surrounded, as they all are, by coral reef and lagoon. They grounded on the reef and stayed there till midnight, when the tide floated them. Then the pilot taxied full at it, got over with floats slightly damaged, and beached the craft. The remainder of the first night was spent in a vain effort to rest in the shelter of the lower planes. In the morning the island was found to be small and uninhabited and without food or water. A sparing meal was made of the emergency ration which both officers carried, and it was decided to try to fly to the main island of the group, keeping in sight of the intermediate ones in case of accidents. Pilot and observer had sacrificed most of their clothing as bandages to protect delicate parts of the engine from the rain. There remained only one charge of the compressed air with which the engines are started, and the pilot used it and got the machine going, forgetting that the surviving garments, such as they were, were drying ashore.

The seaplane rose into the air with its two nude occupants and headed for Male, but the petrol gave out before Male could be reached, and a forced landing was made outside the reef of a fair-sized island. The machine had to be abandoned while pilot and observer swam ashore. They found signs of habitation—a few huts of palm leaves with piles of coconuts, a fire still burning, and a few meagre fowls. One of these they killed and cooked. After another very disturbed night, during which some natives who had approached fled at their friendly greeting of "*Salaam*" they found themselves alone, provided themselves with some kind of clothing, and devoted the day to building a raft with the object of salving the seaplane.

The work was interrupted by a vision of natives in fishing boats making a distant and very gingerly reconnaissance. Signals to them, though evidently seen, elicited no response, and an attempt to swim out to them produced only a hurried retreat. By the evening, however, the natives had mastered their timidity sufficiently to allow the desperate pair to reach them, and they were induced, after much gesticulation, to take the seaplane in tow. It was piloted in through the narrow entrance to the lagoon and stowed on the beach as well as circumstances would allow. This duty performed, the officers surrendered themselves and were taken before the headman.

This personage committed them to a sort of open arrest in which their actions were suspiciously watched without undue interference. They were accommodated in the bachelors' quarters with some fifty or sixty natives. Gradually cordial relations were established. The airmen exhibited their machine and gained a great access of prestige by giving their friends shocks with the wireless battery—a move which ensured for the future no inquisitive tampering. In fact, the "strange and loud-voiced bird," as they called it in their report to the Sultan, was forthwith surrounded by a barrier of ropes. By day they bathed; in the evenings they had sing-songs interspersing ragtime and music-hall tunes with the drowsy chants contributed by native talent. Thus about a week was spent, and finally envoys from the Sultan took them to Male, where they were welcomed and given honorary rank with the correct uniform in the household brigade. This was the kit they were wearing when eventually they disembarked from the small sailing vessel which took them to Colombo.

Now, to this story, which is pretty much what both of the published accounts boil down to, Mr. Kipling, using Guy's letter, made, as I have said, one notable addition. Willie did not tell us, but Guy did, that one night, as an extra turn in the programme of music, Guy induced him to take out his false teeth. In Mr. Kipling's words:

"'He told his observer'"—I like that touch, "he *told* his observer"—"'to show 'em his false teeth, and when he took 'em out the people all bolted.'

"'But that's in Rider Haggard—in *King Solomon's Mines*,' The Damper remarked.

"'P'raps that's what put it into Baxter's head then,' said Duckett. 'Or else,' he suggested warily, 'Baxter wanted to crab his observer's chances with some lady.'

"'Then he was a dashed fool,' The Damper snarled. 'It might have worked the other way. It generally does.'"

We could never induce Guy to tell us whether the motive hinted at by Mr. Kipling was the true one, but from what we knew of him it seemed likely. On the other hand, there is Willie's character to consider. I gather from the editorial note at the heading of his article that his marriage later in the year shook London to the core—"one of the society events of last Autumn" is the exact phrase, but one knows how reticent editors are—so it is perhaps unjust, certainly indiscreet, to suggest that he was trying, however mildly, to cut Guy out. And besides, there is that very straight official report on the native ladies which I quoted above. I think the matter must remain one of the war's mysteries.

LEBANON.

IX

BEIRUT AND DAMASCUS

HITHERTO I have said nothing of Beirut beyond mentioning that it occupies the extremity of the middle prong of our three-pronged-fork system of railways. It was the only coast town of importance which was not included in the comprehensive system of operations at the end of August, 1916, and the omission was due to the fact that the piece of railway joining it with the back-bone line running south from Damascus, which, as I have said, is really the beginning of the Hejaz railway, is only a branch to the port and not part of the lines of communication which it was the object of those operations to harry and reconnoitre.

The line and the town were, however, playing their part in the war none the less, and frequent visits were paid there by the seaplanes at other times during 1916. One of its uses was as a port of embarkation for the coastal traffic which was supplementing the railway supplies for the Sinai front; and during 1916 it came very strongly under suspicion as a possible base for submarines. It was not until the beginning of 1917 that this

AT BEIRUT.

suspicion developed into a certainty. It then became definitely known that German submarines were putting in there to replenish their stores of food and fuel, and the R.N.A.S. had orders to give it a more striking reminder of their existence than was to be conveyed merely by sending machines to fly over the town and surrounding country.

The state of affairs was, of course, not common knowledge. People who knew Beirut before the war—and it had then a mixed population numbering nearly 200,000—may have been surprised when they heard that a place so attractive and peaceful-looking should have become the object of attack, and the statement in an official Turkish report to the effect that, although undefended, the town had been subjected to bombardment, possibly increased the feeling of surprise among those who were not accustomed to read between the lines, or, as perhaps one may put it, to read between the lines. It was, however, very necessary to read between the lines.

A point always to be remembered is that there is the greatest difference between offence and defence. A town like Beirut; which certainly was undefended to the extent that it was not bristling with big guns (though it had its entrenchments and armed forces), may none the less play a part in a very formidable offensive. That is the part which had been forced upon this pleasantly situated, innocent-looking seaport, with its business-like harbour-buildings overlooking the safe anchorage and its attractive residential quarter, nestling at the foot of Lebanon. It had unquestionably become a legitimate target by 1917, even if, in the opinion of our perhaps too punctilious higher command, it had not become so before.

Several times during 1916 and once in February, 1917, seaplanes flew over the town with no more offensive intention than that of ascertaining the nature of its land defences and the positions of camps and concentrations of troops in its neighbourhood. The last of these flights calls for rather more than mere passing mention, because it constituted what I believe was then, at any rate, a record for a seaplane flying overland. The machine actually reached Damascus, which is about fifty-five miles in a straight line from Beirut. If we add the distance between the ship and the shore, and a margin for deviations from the straight line, to say nothing of the Lebanon Range, which is anything from 4,000 to 7,000 feet high, we shall have a total flight out and back of at least 150 miles, of which only about thirty were over water.

The flight was carried out from the Anne. The ship left Port Said on 26 February, 1917, with orders to make a seaplane reconnaissance of the country near Haifa and the Carmel Range, and then to proceed to Beirut and make similar observations across the saddle of Lebanon. In each case special attention was to be paid to the nature and condition of the roads. The idea was, in fact, to gather more definite information than had been hitherto ascertainable as to enemy transport possibilities in view of the British advance in Palestine, which was in gradual progress throughout the year and culminated in the rapid forward movement on the capture of Gaza in November.

The Haifa flights were successfully accomplished on 27 February, and the carrier then sailed north, arriving off Beirut early in the morning of the 28th. A two-seater machine got off the water at 7 a.m., and the ship's company went unconcernedly to breakfast and to their various jobs, expecting to see it back somewhere about half-past eight. By nine they were beginning to get anxious, and when by nine-thirty there was still no sign of the seaplane, there was much speculation in that suppressed half-hesitating manner which people adopt when their feelings fight for expression in spite of themselves.

It was ten minutes to ten when the machine was at length sighted high in the distance. Before the pilot and observer got on board, anxiety had given way to knowing explanations of the delay. Of course the old 'bus had let them down. The engine had conked out—everybody (excepting the Engineer Officer) knew it would. Or the water had boiled; or frozen—the suggestions were not all serious—getting over those mountains. Or this, or that, or the other. But they were all wrong.

The delay, of course, had nothing whatever to do with engine trouble. It had, in fact, a good deal to do with precisely the opposite. The pilot—Noisy Dan, we used to call him, because he was the quietest person who ever gave a word of command—having taken the machine in as far as Rayak, forty-five miles or so from the ship, and found that it was running well, just went on to Damascus in spite of signals from his observer, who thought he had mistaken the direction of home. But he was not making any mistake. He had that instinct which some pilots have (though it sometimes plays them false) for the exact extent of the demands they can make on their engines. Theoretically it was a risky thing to do, and it would never have been ordered beforehand. But Dan happened to be in command of the operations. He had a feeling that what he wanted to do could be done, and he did it.

The reconnaissance from Beirut to Rayak took in the middle prong of our three-toothed comb, and at the risk of being tedious, I must say another word or two about it. I have spoken for the sake of convenience of a continuous system of railway from Aleppo through Damascus to Medina in Arabia, but technically it is not continuous. There is a break at Rayak. North of Rayak the line is of the normal continental gauge; south of it the gauge is narrow. There was therefore no through traffic. At Rayak, for travellers and stores between the north and

BEIRUT AND DAMASCUS

the south, it was a case of " all change." There was consequently an elaborate system of transhipment platforms where the broad and narrow gauges ran parallel. A further point of interest is that the very steep gradients crossing the Lebanon range between Beirut and Rayak necessitate the use of a rack and pinion system with engines which can work either with or without the central rail and cog-wheels.

The flight to Damascus followed the railway the whole way out, and the machine reached its highest elevation at Zahle, where it flew at 7,600 feet, though Zahle is not the highest point on the railway. It was, however, near enough to the original objective, Rayak, to make a safe altitude advisable. Rayak was inspected from 7,000 feet, and several photographs were taken showing a most elaborate arrangement of railway workshops and sidings. Notes were made of the rolling stock, which was very plentiful, of tents and of such evidences of the practical use of the workshops as smoking chimneys.

Then the journey was continued in the direction of Damascus. About two miles from the town the white course of the highway revealed what to the R.N.A.S. observers in these parts was a good deal of a novelty. There was a battalion of troops on the march. One might suppose that there was nothing so very unusual in this in a land in which there were everywhere signs of war. Trenches, gun emplacements, camps, were common objects of

DAMASCUS: THE STREET CALLED STRAIGHT.

the country-side, but these were fixed things which could not hurriedly take cover at an alarm.

With moving bodies of troops it was different. Seaplanes seldom approached the Syrian coast without awakening into life certain smoke signals which conveyed the news to watchers on the lookout inland, and there was always time for mobile bodies to find some place of concealment. But neither Rayak nor the roads leading to Damascus had ever before been visited by hostile aircraft, and doubtless the warning signals which had produced the utmost alertness some miles in from Beirut had stopped far short of this distant spot where the highway crossed the bare open slopes of the low hills. At any rate, the compact little parties of soldiers, with their accoutrements gleaming in the morning sunlight, and the straggling tail of baggage animals at their rear, were evidently taken by surprise.

Possibly the machine was mistaken for a friendly one arriving for the training school which was about to be started in Damascus, a conceivable error, since at 7,000 feet the difference between a land machine with its wheeled under-carriage and a seaplane with its floats might not be readily distinguishable by inexpert eyes. But the mistake was quickly discovered, and the troops got five or six guns into action with remarkable speed. Their shells flew wide, however, and the machine continued its flight uninterrupted, and in another minute or two it was hovering above one of the oldest cities in the world.

Damascus has been called by the Arabs "The Pearl set in Emeralds," and to see it from the air with its pale mass of houses lying in the midst of its luxury of trees and fields, so I learned from the lucky observer, is to realise something of the truth and beauty of this description, though it applies rather to the view of it which, according to tradition, Abraham, and after him Mohammed, had from the Jebel Kasiun, the hill lying two or three miles to the north-west. The town, which may resemble a single pearl when seen from the ground level of the hill, 2,000 feet and more above the level of the plain, has, when seen from the air, more the appearance of many pearls thrown down into a heap, with the edges spreading away irregularly and in small clusters. And one may imagine, perhaps, that the "Street which is called Straight" is a line drawn through the heap by the jeweller seeking with the point of his calipers for the biggest or finest specimens of the collection, a fancy which need not be affected by the fact that part of the street is arched with corrugated iron.

But neither the pilot nor the observer had much leisure for such thoughts as these. They were there to make as close an inspection as time allowed, not of pearls or emeralds, but of trenches and gun-pits, camps and hangars. The machine circled over the town and the plain, and, the necessary notes being made, set off again on the return journey. This time the main road was followed in preference to the railway, which separates from it seven miles from the city, and the reconnaissance of this showed another battalion of troops and several smaller parties all marching in the same direction. Their faces could be seen as they gazed upwards in wonder and bewilderment, but few of them were ready with their weapons, and no damage was sustained from the very spasmodic fire which they put up. The saddle of Lebanon was crossed once more at Zahle, where a photograph was obtained of a prospect of snow-clad peaks with the white drift clinging in the hollows and filling the zig-zag grooves of the watercourses. This photograph, the last of the series taken, had no military value excepting possibly that it gave negative evidence. It showed no trenches, no fortified posts, no guns, no men; but I think when the authorities saw it they forgave the waste of a plate.

It was soon after the return of the Anne from this trip that Beirut was promoted—or degraded—from the list of places to be reconnoitred to that of places to be attacked. During 1917 the R.N.A.S. bombed it on three different occasions, and their attentions were graduated in intensity.

The first operation, carried out on 13 May, was in the nature of a gentle hint. The Empress had come out to take the place of the Ben-my-Chree, and this was one of her first trips. She arrived off

BEIRUT HARBOUR, WITH TURKISH GUNBOAT SUNK DURING ITALIAN-TURKISH WAR, 1911.

BEIRUT AND DAMASCUS

the coast early in the morning in time to see the dawn come up behind the snow-clad peaks of Lebanon, and to wonder at the curious effect of the ground swell rolling the Mediterranean into long interlacing streaks of blue reflected from the sky above and dazzling orange reflected from the distant sunrise. Beirut, as seen from the air, presents a big, spreading, huddled mass of houses on a broad plain at the foot of the Lebanon range. The town has its little hills and dales, and the buildings, many of them the typical flat-roofed, square native dwellings, many more the gabled and variegated structures of European design and occupation, hug the sides of the slopes and crowd each other in a confusion which seems hardly lessened by the frequent intersections of narrow streets. All this is in the neighbourhood of the harbour, which forms a triangle with the entrance between a short arm running north and a long one running about north-east. Inland in all directions from the harbour the houses spread out with more breathing space, until they appear dotted among pleasant fields or crouching above tiny inlets of the coast, forming delightful-looking residential suburbs with tramways threading the broad roads that give access to them. The only sign of war was a sunken ship in the middle of the harbour. But this was almost rather a sign of peace, for it was a Turkish gunboat which, with another wreck, had lain there undisturbed since the days of the Ital-

ABOVE AND BEYOND PALESTINE

ian-Turkish War of 1911.

On this sunny May morning three seaplanes flew over the harbour, but only five bombs were dropped. One of these, however, was a 500-lb one, and in order to carry it, the machine, which was a double-seater, had to dispense with its observer. The bombs from the other two machines fell close to the mouth of the harbour near the place where the submarines were accustomed to moor when they were loading stores. The 500-pounder fell short. It dropped, in fact, into a piece of open ground which lies just west of the base of the triangular harbour. The officer who dropped it confessed afterwards that it seemed hard luck to launch so much destruction at a place which reminded him of a peaceful English seaside town on a Sunday morning—

> " Thus the doing savoured of disrelish:
> Thus achievement lacked a gracious somewhat."

I don't suppose he had Browning in his mind, but anyhow, the bomb fell short. If it was regrettable that it should not have hit the mark for which it was intended, it was certainly lucky that it didn't fall on dwelling-houses. The crater made by its explosion was 15 feet deep and 24 feet across, and it not only made this very deep impression on the face of the land, but it made also a very deep impression on the minds of the population. We heard afterwards that nobody in Beirut imagined

BEIRUT AND DAMASCUS

that a bomb dropped from an aeroplane could make such a hole. They asked each other " how the Turkish Government could strive against a people who used such infernal machines."

This, as I said, is what we called a gentle hint. But just as there are certain good folk who cannot see a joke without a preliminary surgical operation,

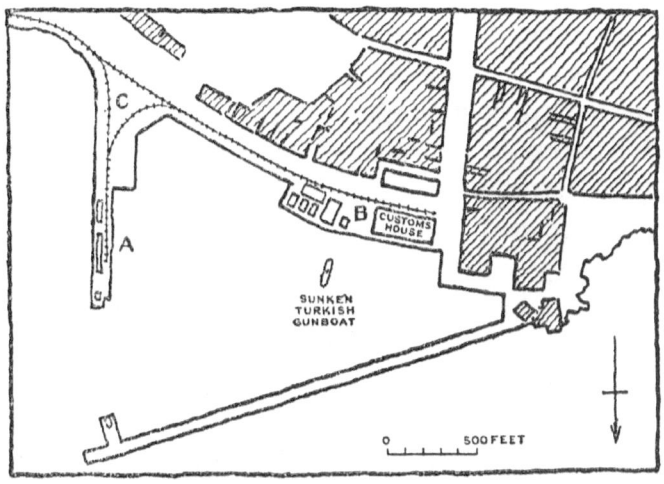

SKETCH MAP OF BEIRUT HARBOUR, SHOWING
OBJECTIVES A, B AND C.

so there are others who cannot take a hint even if you drop it at their feet with 500 lb. of high explosive. The Germans, who had forced upon Beirut the unwelcome character of supply base for their submarines, showed themselves to be among this unimaginative class. They didn't take the hint. Submarines continued to use the harbour, and the required supplies were still taken aboard, though possibly it was done with greater precaution than previously.

So a second visit of the Empress was arranged, and these things were taken into account. This was on 17 August, 1917. Four seaplanes attacked and dropped no fewer than eighty-four bombs, though they were not 500-pounders. A photograph taken during the May expedition was used by pilots and observers as a sort of flying map, and the spots to be attacked were carefully marked on it. It showed the triangular harbour, of which the short arm with its long, narrow sheds was called objective " A " (see sketch map, p. 195). These sheds were believed to contain naval stores, because it was definitely known that the submarines, when they entered the harbour, were moored alongside this projecting arm. Objective " B " included the buildings on the foreshore at the town side of the harbour. All these buildings were sheds or offices connected either with the harbour or with the railway, which had lines running parallel to the water's edge and along the short arm known to us as the eastern mole.

There was a third objective, " C, " the little piece of ground from which the eastern mole, with objective " A, " springs from the mainland. This was regarded as a possible storing place for petrol, and it had in fact been actually so employed, but on the occasion of this attack it appeared to contain nothing of military value, and as it was very near a group of dwelling-houses, it was decided not to take the risk of bombing it. This decision, it may be noted, was made in view of explicit orders that

ENTRANCE TO BEIRUT HARBOUR.
The ships in the distance are moored against the mole called objective "A" in the operations.

no building or part of the town was to be molested unless it was known to be engaged directly or indirectly in assisting the work of the submarines.

The four machines were in the air soon after 4.30 in the morning, and the leading one began the operations by firing two trays from its Lewis gun at the extremity of the long arm of the harbour where there was a watch-tower. That the attack was a surprise seemed evident, because it was only after the machines had been engaged for some time that a flag, doubtless intended as a warning to ships not to approach, was seen to have been hoisted on the tower. The first and second machines had orders to look out for submarines, and, if any were seen, to descend to a low altitude and attack them. Failing this, they were to reserve their armament, take photographic records of the work of the other two, and when it was finished to discharge their bombs at whichever of the objectives seemed to have suffered least.

In the event there was very little choice. Numerous direct hits were obtained on both the objectives "A" and "B." As regards "A," bombs dropped through the roof of the shed exploded inside and started a fire, railway trucks were hit and overturned, and piles of stores lying at the end of the mole were scattered and ignited. With "B" more was done, but "B" was a bigger target and offered greater scope. Numerous hits were obtained on the sheds at the water's edge, causing internal explosions and fires which were blazing fiercely before the machines came away. In the largest of these sheds there was hardly a whole tile remaining on the roof, and through the gaps the flames could be plainly observed inside. Similar results were obtained in the smaller adjacent buildings.

Particularly good work was also done on the big substantial building lying behind these foreshore sheds. The inclusion of the building in objective " B," as with objective " C," had been left to the discretion of the bombers. It was under suspicion, but failing definite evidence of military uses, it was to be given the benefit of the doubt. When the machines arrived, evidence was immediately forthcoming; there was a sentry box on the pavement in front of it. This was good enough, and it received its share of bombs. That no mistake had been made appeared afterwards. The building, which before the war had been occupied by a big general provision dealer, had been commandeered by the Turkish Government for the use of the army, and it paid the penalty with a gaping wound caused by a bomb, which fell plumb on the ridge of the roof, and by many scars and much shattering of doors and windows in its civilian-like face.

Before the operation was finished, the whole of the harbour was veiled with smoke from " B " objective, but in spite of this some surprisingly good photographs were taken. Some of them were given wide publicity in the London illustrated papers. Not a singles bomb fell in any part of the residential quarters, and it may be noted with interest that the inhabitants of Beirut evidently realised that there was little danger in the attack excepting at points of military importance, for many of them were assembled on the roofs and verandahs of their houses watching the proceedings. By ten minutes to seven the last machine had returned to the ship.

Among the photographs taken during the flights over Beirut there was one which aroused a certain amount of excitement at the Base. With the aid of a magnifying glass, the observer, V., who had taken it, discovered on the pier used by the submarines a group of objects which he was certain were mines. Further examination of other photographs with a more powerful magnifying glass showed in the harbour what was unmistakably a dredging barge, and the " mines," which were undoubtedly of a rather unusual shape, became quite intelligible if they were regarded as spare dredging buckets. This they turned out to be, a fact which I was able personally to verify soon after the Armistice. But V., who was not easily persuaded, remained unconvinced to the end.

A few weeks after the second visit information from naval patrols again made it clear that the harbour was still being used as a submarine supply base, and more elaborate plans were laid. Whereas the second attack had occupied barely an hour and a half, the third, which took place on 27 September, began at about a quarter to six in the morning and was not over till nearly twelve. As before, the objectives were the short breakwater and the buildings and store-sheds facing the anchorage, but on this occasion the bombing seaplanes concentrated their attention only on "B," the second of these objectives; while "A" was bombarded by the guns of one of H.M. ships, with seaplanes observing and directing the fire. This division of labour, it may be noted, arose from the fact that "B" objective could not have been attacked by naval guns without the risk of destroying the surrounding houses or property of civilians. The battleship therefore took up a position from which the mole could be attacked by shells which would pass across the front of the town, and failing hits, fall into water, whether they were short or over.

The seaplanes opened the operations by flying in over the harbour soon after five in the morning. The first thing that struck the observers, who were naturally interested in having another look at the traces of their previous bomb-dropping, was that the Turks had evidently been very hard at work. Some of the sheds which had been completely stripped of tiles had now been reroofed, and others were undergoing the process. The hole in the roof of the big building at the back had been repaired, and broken glass had been renewed. All this, which was plainly visible at the low altitude—800 feet or so—at which the machines were flying, pointed unmistakably to the importance attaching to these objectives in the Turks' estimation, and it was with much satisfaction that when the morning's bombing was finished most of the work was left in a condition to be started all over again.

The naval guns began their share of the programme just before ten o'clock, and continued for an hour and a half. The fire was accurate and numerous hits were recorded. This terminated the third and last attack on Beirut, and this time the Turks or Germans seemed to understand that we were in earnest. In the military and naval language of high explosives, we had stated three times the plain fact that we did not approve of the use of Beirut as a civilian mask for military or naval enterprises. On the first occasion, and on the second, this plain statement was apparently not believed, but on the third it was accepted. As the Bellman said, "What I tell you three times is true." Submarines did not entirely discontinue their use of the base, but their visits were hence-forward far less frequent and far more circumspect. To use a service technicality, "They had got the wind up."

X

GAZA AND JERUSALEM

THROUGHOUT 1916 and 1917 the British Army had been gradually pushing its front from the line which it occupied when the Turks made their second attempt to invade Egypt. One by one the desert posts in Sinai were won by fighting or occupied after evacuation by the enemy in the face of the steady improvement in our lines of communication. Into the details of this advance, with the hardships of desert life as a constant accompaniment to the persistent efforts of a foe by no means despicable, it is not my purpose to go. The part played by the R.N.A.S., though it was, as I believe, of the highest importance, was always an indirect one. We have seen how they kept watch over the inundated tracts which formed part of the defences of the Canal, and we have seen that their reconnaissance and bombing raids were directed against numerous positions and districts of country which at the time the flights were made were out of reach of the R.F.C. The work of the two forces seldom overlapped. There were one or two occasions when they worked in conjunction, notably the attack, already mentioned, on Tul Keram, on 23 June, 1917, when seaplanes bombed the railway junction while land machines occupied the attention of the Turkish aerodrome at Ramleh. But such combined operations were the exception. As the Army advanced, the Naval Air Service advanced too, or, rather, it did not so much advance as compress its area of enterprise, for from first to last, all available points behind the Turkish front right up to Adana and the approaches to the Taurus tunnel were included within the field of its investigations. The forward march of the British troops merely meant to the R.N.A.S. that certain places which they formerly visited, photographed, and disfigured with high explosives, were transferred to the very efficient care of the R.F.C., who carried on the good work.

In this way, by the beginning of 1917, every part of the Palestine coast as far north as Jaffa had dropped out of the R.N.A.S. list. El Arish had come into British hands on 22 December, 1916, Rafa on 9 January, 1917, and Khan Yunus, fourteen miles south of Gaza, by the end of February. North of Jaffa was still our province, and Tul Keram, Haifa, Beirut, Chikaldere, and Adana, were all inspected and—dealt with on one or more occasions. With the exception of the attacks on the submarine supply-base at Beirut, these expeditions were always carried out with the object of assisting the plans of the Army which was faced with the problem of capturing Gaza and Jerusalem.

GAZA AND JERUSALEM

In March, 1917, the Turkish front extended from Gaza to the south of the Wadi Ghuzze and thence on to Beersheba. It was not a continuous line in the sense in which one thinks of the Western front as a continuous line. Gaza, on the enemy's right, with the Jebel el-Muntâh, the hill to which Samson once carried the city's gates, a prominent feature, was strongly defended, as was Beersheba, though in a lesser degree, on the extreme left; and there was a well-fortified position in the centre at Sharia. Between these points the connection was by means of patrols rather than by material defensive works, and there was need for constant observation by the R.F.C. of much the same character as that which the R.N.A.S. were making over the same ground a year earlier.

It was this position in respect of Gaza and the front generally that General Murray determined to attack at the end of March, 1917. The Wadi Ghuzze was crossed on the 26th, but the fight lasted two days and supplies, water, and ammunition could not be brought up in sufficient quantities. Gaza was not taken, and our troops retired again across the Wadi on the 28th. A further effort was made three weeks later, when the railhead had been pushed up to Deir el Belah and transport facilities had been greatly improved.

During the second attack, which lasted from 17 to 19 April, the Empress played a small part.

French and British battleships were conducting

a coastal bombardment, and the seaplane carrier executed submarine patrols until the machine—they had only one on board—was put out of action through the breaking of a float on landing. Having nothing else to do, we watched the battle through our glasses from the deck or played cards in the wardroom. There was only one thrilling interruption. High in the air a distant aeroplane was sighted, and as we stood on the bridge it seemed to be making a line straight for us. Was it an R.A.F.

H.M. LANDSHIP WAR BABY,
LOST DURING THE ATTACKS ON GAZA.

machine, or were we to be bombed by a German? This question was not decided until the airman wheeled landwards as he reached a point almost above us, and we saw that his wings bore not black crosses, but coloured circles.

The Empress returned to the Base on the 18th. The second attack on Gaza also failed, and the respective positions of the two armies remained practically unchanged for the next six months.

During this period both sides made ceaseless

preparations for the inevitable struggle. The front developed from a line of detached posts to a series of trench systems with very few gaps. The Turks constructed light railways and roads available for motor transport, and provided arteries by which ammunition and reinforcements could be conveyed readily to any point of their front. This they fortified with all the ingenuity gained from three years' experience of war. By far the greatest strength in their line, so far as entrenchments and batteries are concerned, was concentrated round Gaza, and the reason for this was the perfectly sound assumption that any attempted advance by the British must keep touch with the sea coast, since the country inland, with its waterless desert, presented almost insuperable difficulties of supply. The Turks and Germans, moreover, as we have already seen, had always been rather in fear of attack by sea all up the Syrian coast. Gaza was accordingly fortified with a wide scheme of trenches which were regarded as absolutely impregnable. An officer prisoner taken before the advance ridiculed the idea of capture, and there was some excuse for his confidence for it was found afterwards, when the position was captured, that there were many dugouts with head-cover 9 feet thick, and winding stairs leading to shelter a dozen feet underground. This was the condition of the Turkish right.

Their left, round Beersheba, was pretty well defended too, but as they rightly guessed, Gaza was the place we wanted, and there the real strength lay. General Allenby, who had taken over General Murray's command on 28 June, 1917, faced the problem by what was in effect a masterly combination of strategy, tactics, and psychology. He employed, as all good generals must, the process which Edgar Allan Poe in one of his detective stories described as "an identification of the reasoner's intellect with that of his opponent." He thought not only what the enemy was thinking, but he thought a step or two ahead.

He began his advance on Gaza by a reconnaissance in force on Beersheba, and he knew exactly what the Turks would say about it. They would say: "This is a clever move to distract our attention from the place where the main attack is coming off. He wants to weaken our hold of Gaza. He is not really going to waste his time over Beersheba." They did not, therefore, hurry reserve troops to the position; they relied on the considerable force that was already there and the natural defence comprised in the waterless desert which spread for miles on their extreme left. Here, for the sake of further security, they sent out armed patrols, some of which came in contact with our cavalry, but they did not allow themselves to be deluded by what they thought was a transparent ruse.

Realising, as I picture it, that the Turks were arguing in this way, General Allenby began a terrific bombardment of Gaza, one of those bombardments which were well known on the Western front, but a new thing in Palestine. "There, you see," said the Turks, "we were quite right. Here he is knocking his head up against the impregnable stronghold of Gaza, just as we said he would," and they sat tight in their 12-foot dugouts and replied to the bombardment with counter bombardments and overhead reconnaissance in the firm conviction that the attack, when it did come, would waste itself and recoil as the two former attacks had done earlier in the year.

But in the meantime, the demonstration at Beersheba, twenty-five miles off, had developed into something more serious. The reconnaissance in force had withdrawn, and nothing apparently was coming of it, when early in the morning of 31 October our infantry after a forced night march, and our cavalry after a wide sweep round the south and south east—some of them rode thirty miles before getting into action—captured Beersheba and smashed the extreme end of the Turks' entrenched line. It was not done without very hard fighting, and there were incidents which recall the supreme valour of Balaclava, but the move was an outstanding success, and it was the beginning of the end of Gaza.

The commander of the Turkish forces, the Bavarian, Kress von Kressenstein, who certainly worked with a considerable amount of ingenuity, though it was insufficient, conceived at this point the idea

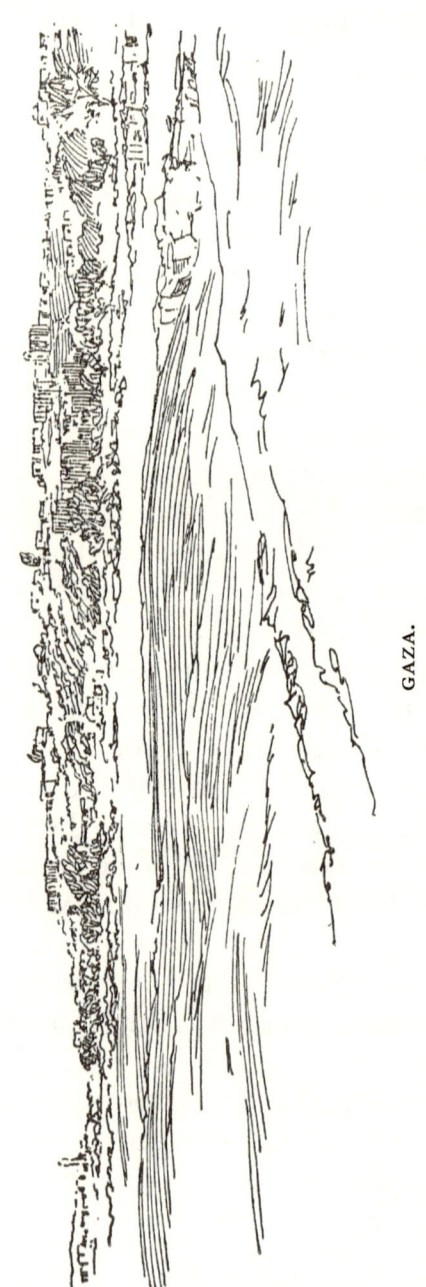

GAZA.

GAZA AND JERUSALEM

of luring the British right flank into a pursuit of the retreating Turks in the direction of Jerusalem to the north-east. This way lay hilly and difficult country in which rear-guard actions could be fought with the greatest advantage. But the trap was perceived. The captors of Beersheba, advancing north in their own time, occupied the heights of Khuweilfeh on 11 November, and remained there nearly a month. Meanwhile they fanned out to the north-west, threatening the Turkish central communications, and ultimately those of Gaza itself. On 7 November the town was taken at the cost of a few casualties, and the whole Turkish Army was in retreat.

In the scheme of things sketched in this brief survey, we can now indicate the part played by the R.N.A.S. Throughout the action the illusion had been maintained by means of an incessant bombardment, that there was to be a direct attack on Gaza. But if this bombardment, accompanied as it was by a steady form of pressure was, strictly speaking, merely a ruse to divert attention from the turning movement on the British right, the ultimate end of it was the same as if it had been the prelude to the advance of strong storming parties. When an army pumps shells into a position without a pause for a week, it is a natural assumption that he has some designs on that position. As Mr. Bramah wisely remarks in *The Wallet of Kai Kung*, "When struck by a thunderbolt it

ABOVE AND BEYOND PALESTINE

is unnecessary to consult the Book of Dates as to the precise meaning of the omen." The precise meaning in this case was that the Turk had to go. Whether he was pushed out from the front or compelled to depart because things taking place at his back made it unsafe to remain, was immaterial to the main issue. He had to go. And as there were certain definite roads and railways and bridges along which he was compelled to travel when he did take his departure, and along which he was bringing up supplies until the necessity for departure had been borne in upon him, it became advisable to drop thunderbolts there too. The omen had to be made quite clear.

The town of Gaza is about a mile and a half wide, measuring to its extreme outskirts, and there is about a mile and a half between it and the sea. The Turks had constructed a branch railway from their main Beersheba line, and this and the road close to it, three or four miles from the coast, provided what was in effect their only way of escape. The railhead was at Beit Hanûn, a village about four miles along the road from Gaza, and four miles from the coast. Two and a half miles farther on was the village of Deir Sineid, where the road crossed the Wadi el Hesi by a bridge, and where another bridge was in course of construction for the railway, which in the meantime was carried over the river bed, then dry, on an embankment. At Deir Sineid, too, there was a secondary branch line running to Huj, the base of supplies for the Turkish tight centre. It was at this most vital point of the enemy's communications that the seaplanes had their earliest glimpse of the Gaza struggle.

The proximity of the road and railway to the coast made it possible to extend the Gaza cannonade very much farther back than would have been practicable by means of land artillery. The

BRIDGE AT DEIR SINEID.

bombardment of these lines of retreat was undertaken by battleships which, lying fairly close inshore, could range on them with the greatest ease, though their targets were generally invisible. It was here that the seaplanes came in. They were employed in spotting; that is to say, they circled above the targets, and the observers, noting where the shots fell, corrected the aim of the gunners by wireless signals.

The City of Oxford, who had recently joined the Squadron, was the first of the seaplane carriers to take a hand, and she arrived at a position off the mouth of the Wadi el Hesi at 9.30 in the morning of 30 October, the eve of the capture of Beersheba. Two seaplanes had, however, already reached the scene of operations, though they had beaten her by only a few hours. The two machines, with a pilot and observer, and the necessary mechanics, had been taken on board a monitor which fetched up about dawn of the same day. One of these machines got to work at 10, flying over the sand dunes to Deir Sineid to spot for the battleship's 6-inch and 14-inch guns.

Fire was opened on the railway station, and several direct hits had been scored before the observer noted a large dump by the side of the line near it. It had evidently been placed there for transportation by means of the branch line to Huj. It was impossible to make out whether the dump consisted of ammunition or of other stores less inflammable, but in any case it provided a good target. So a signal was sent and the 6-inch gun was ranged upon it. Half a dozen or so test shots were fired, and then there was a direct hit. The dump made an immediate and gratifying response by proclaiming itself to be ammunition. It exploded and continued to explode for thirty-five minutes, demolishing the railway station and tearing up many yards of line. It was one of the occasions on which an observer finds supreme satisfaction in his work, and the machine circled round watching the sight for a few minutes before returning to the ship.

In the afternoon the same pilot and observer experienced a thrill of a different kind. They were over the same ground and were about to return home when their machine was attacked by a single-seater Halberstadt Scout, one of the aeroplanes with which the Germans had managed to put some backbone into the Turkish flying service. The estimated speed of the enemy was nearly double that of the seaplane, and the only course open to our pilot was to make hard for the sea in the hope of bringing the Scout within range of some of the ship's guns. The enemy machine, however, followed so close that anti-aircraft fire was impossible without endangering the friend equally with the foe, and it was not until both had descended to 800 feet, and the seaplane, having used its Lewis gun at close range, had broken away with a sharp turn aside, that the monitor was able to open an attack and drive off the pursuer. The seaplane was found to have been hit in thirty-five places, and one elevator control was shot away. The pilot escaped untouched, and the observer had a slight splinter wound which interfered only temporarily with his work.

The naval bombardment of these and other points on the Turkish line of communication was continued for several days, with constant seaplane co-operation which was maintained not without further interference from enemy aircraft. Machines and ships were attacked from the air, but

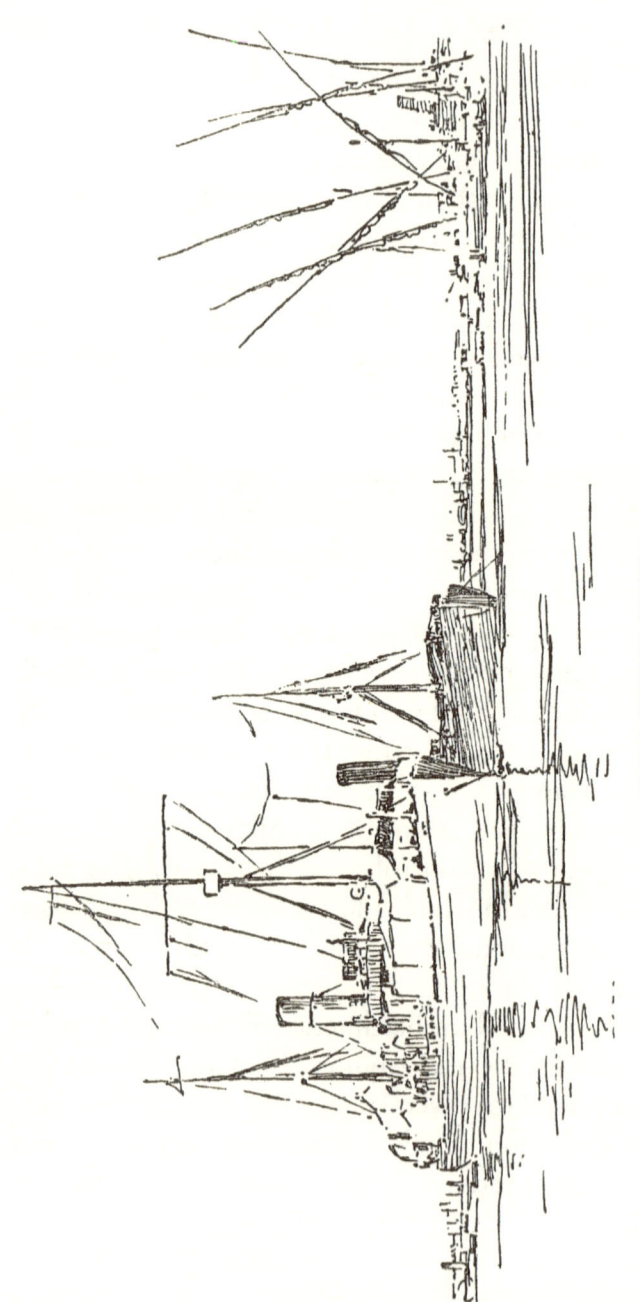

H.M.S. CITY OF OXFORD.

the Halberstadt effort of the first day was the most daring and effective. On 4 and 5 November, seaplanes both from the City of Oxford and from the monitor were spotting over El Nasle, a position just north of Gaza where a Turkish battery was impeding the advance of our left, which was making steady progress. On 6 November information was received from the Army that the enemy was retreating northwards through Deir Sineid, and that the traffic of convoys there was becoming congested. The City of Oxford accordingly proceeded again to the mouth of the Wadi el Hesi, and throughout the day seaplanes were engaged in spotting for the guns of the French cruiser Requin.

On 7 November Gaza was in our hands and the Turks were making desperate efforts to check our advance and cover their own retirement northwards, by occupying a line of trenches previously prepared south of Askalon, protected by field guns between that town and Deir Sineid. The Requin and the City of Oxford followed up the coast, and seaplanes again directed the fire of the French guns on these targets. In the course of the morning the French cruiser was joined by a British cruiser and a monitor, and their fire was directed at Julis, a station on the line to the north-east of Askalon. Here there was reported to be another big accumulation of Turkish transport, and it was the last point on the line of retreat which naval guns could hope to reach with any accuracy. The spotting was carried out under difficulties. There were low clouds, which compelled the machines, if the observers were to see anything at all, to fly at below 1,200 feet.

During the evening of this day British cavalry and guns had been pushed along the coast to the mouth of the Wadi el Hesi, and although the Wadi was occupied by the Turks about five miles inland, our troops continued to advance along the sea-shore, where they were threatened by shrapnel from enemy guns situated over the sandhills on their right. All ships thereupon took up firing positions off the coast between the Wadi el Hesi and a point a few miles north of Askalon, with a view to driving the Turks inland. On 8 November seaplanes spotted for the monitors and the French cruiser, and by mid-day our left flank had advanced well on towards Askalon. In the afternoon the City of Oxford left the scene of operations, and was back at her base the next morning.

In the meantime, while the operations were in progress off the Gaza coast, the Empress had been assisting in the general scheme at points farther afield. About twenty-three miles north-east of Jaffa the railway crosses at Jiljulie, a branch of the Nahr el-Auja, a river which reaches the coast about four miles north of the ancient port. The destruction of this bridge, which was not on any of the Turkish branch lines constructed for the defence of the Gaza front, but on the original main line running to Beersheba, was planned with a view to impeding the transport of supplies to the whole front, the attack on it by machines from the Empress formed part of a scheme of operations which included also the bombing of oil factories at Haifa, and the railway and rolling stock at Tul Keram. The idea was to carry out the whole of the flights with small single-seater machines, but the proceedings were attended with a quite unusual degree of bad luck. Six machines of two different types were detailed for the Jiljulie flight, but on being hoisted out, one of them immediately began to sink by the tail. In a few minutes it had turned over and was completely water-logged. Efforts to salve it failed owing to the danger of the bombs which it carried exploding, and it had to be sunk by gunfire. In view of this disaster, due in a large measure to the choppy state of the sea, the two other machines of this type were withdrawn from the operation, and three only proceeded inland. Arrived at their target, they found very adverse atmospheric conditions. Not only was the flying very "bumpy," but the wind was in precisely the wrong direction for accurate bomb dropping on a long, narrow mark such as a bridge. The bridge was not destroyed, though much damage was done to the line and embankment near it.

Having taken the machines on board again, the Empress proceeded to Haifa, where five seaplanes were hoisted out early in the afternoon. The first machine had orders to drop two small bombs in the vicinity of the factory in order to give the employees time to evacuate and take cover. The others, after an interval, scored four direct hits. Three machines only returned to the ship. The first of these reported that two machines had been seen on the water in Haifa Bay. A signal was accordingly sent to the French destroyer acting as escort, and she immediately made at full speed for the harbour. It was not until 6.30 in the evening that the Empress got a signal from her: " All saved, machine lost."

It appeared that the engine of one of the two machines had suddenly " cut out," with the usual unpleasant lack of adequate notice. The pilot was able to make a safe landing so far as airmanship was concerned, but a very unsafe one as regards his proximity to the hostile town, which was hardly a mile away.

The pilot of the other machine saw what had occurred, and after dropping his bombs came down alongside his companion. Rifle bullets from the town splashed round them in the water, but before swimming to the rescuing machine and abandoning his own, the first pilot made an attempt to set fire to it. Having nothing but matches, and not having passed the scouts' test of lighting a bonfire with one match only, he had to give it up, reluctantly leaving behind, not only the seaplane, but also the empty match-box. He reached the other machine and boarded it, but it was found that the extra weight on the floats immersed them so deeply, that the spray thrown up by them broke the propeller. There was then nothing for it but to wait and watch whether a capturing party from the town would arrive before the rescuing French destroyer. Possibly the towns-people were afraid of machine guns, for they allowed the destroyer to win the race, rescue the two stranded airmen, and destroy one of the machines by gunfire. When they reached the sea-plane carrier it was too late to start for the third part of the day's programme, the attack on

TUL KERAM.

Tul Keram, and the Empress put back to Port Said.

With the return of the City of Oxford on 9 November, the share of the R.N.A.S. in the Gaza operations was over. The country over which the seaplanes had been spotting was in our hands before many days had elapsed. Ramleh and Ludd, familiar to the seaplanes of 1916, were occupied on 15 November, and Jaffa was entered without resistance on the following day. The Army pushed on a few miles to the banks of the Auja river, and there planted its left flank in sight of the ruins of a crusader castle at Ras el-Ain.

It was then time for a move north of Beersheba. The troops at Khuweilfeh, who had occupied that place since they took it on 11 November, captured Hebron on 6 December, and advanced ten miles to the north. Jerusalem surrendered on 9 December.

With these operations the R.N.A.S. was, of course, not directly concerned, but the definite halting point reached by the Army suggested that the moment had arrived when there might reasonably be a lull in the routine of the Naval Air Service. The welcome order went round that all officers in rotation could take four days' leave in Cairo.

RAS EL-AIN.

XI

ON LEAVE

No one on active service ever gets away on leave from his unit entirely free from the possibility of being summoned back at a moment's notice. That is one of the understood conditions, but on the whole, I suppose the right of recall is exercised only in a small minority of cases. There is generally some one else to " carry on." But with the R.N.A.S. at Port Said there was very seldom this accommodating surplus, and the flying officers—pilots and observers—had nearly always the disturbing satisfaction of knowing that they could not be spared. Leave was not easy to get, and when obtained, it was unusually liable to abrupt curtailment at the most unwelcome moment.

A grant of four days' leave in Cairo to all officers in rotation two by two—such was the order—was, therefore, a thing so unprecedented that one might reasonably assume that those in authority could foresee a quiet time ahead. The assumption, as a matter of fact, was not correct, because there was renewed activity long before the leave list had been exhausted, and in one case at least a telegram

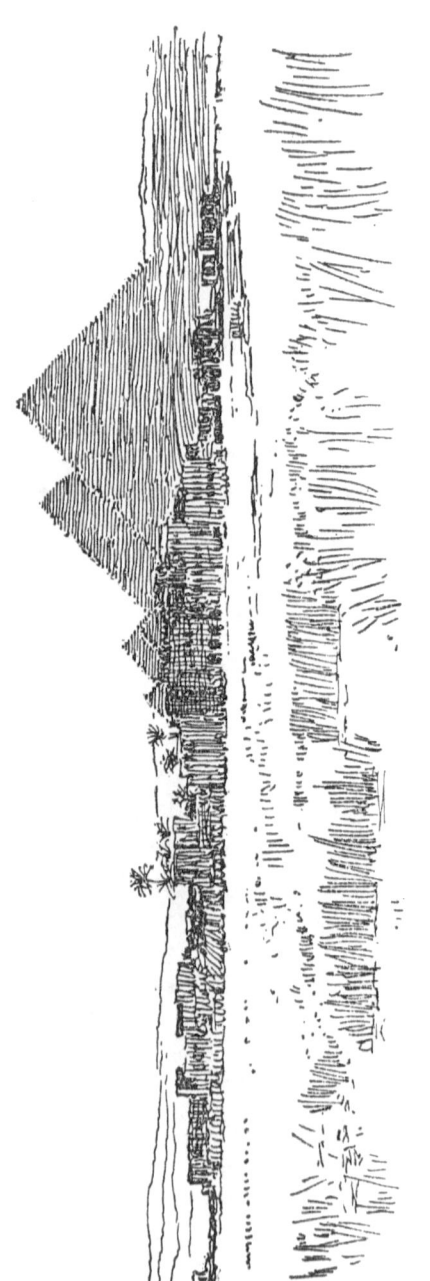

EL-TALIBIYEH AND THE PYRAMIDS.

wiped out the fourth day. But the assumption at any rate was made, and it dispelled the "now or never" sort of feeling which, as a rule, allowed no time to be wasted between the granting of leave and its enjoyment. There was, indeed, on this occasion, a very noticeable hanging back. Officers vied with each other in generous endeavours to induce some one else to go first. The fact was that by putting the period off it might be made to include Christmas Day, with its cheery hotel festivities.

Not that Cairo needed any extra attraction. It is hardly necessary to say that to get to Cairo was one of the ambitions of everybody who came to Egypt. The desire to see the Turk well away from the threshold of the territory which, by deputy, he had so long mishandled was, officially, of course uppermost, but this desire never entirely overshadowed that of seeing something of the territory itself. We were all at heart tourists, and "a drop of leaf" gave one the excuse frankly to appear in that character. But all the same, the chance of going to Cairo at Christmas made it worth while to wait a week or two longer, and there were a good many candidates for the four days which fell round 25 December.

The matter was settled amicably enough. One of us at any rate was admitted by general consent to have been indicated by fate for the privilege. This was Puck, who the year before had been ripped untimely from the Continental Hotel on Christmas Eve and ordered back to Port Said to take charge of the flight which went in the Raven to bomb Chikaldere Bridge. Among the others there was much sorting out, but such things arrange themselves. The spin of a coin has decided many more important matters.

Different people went to Cairo for different reasons apart from the business of seeing the Pyramids and the native bazaars, and, for the elect, that of celebrating Christmas. Puck, for instance, had a very clear notion of participating to the utmost in everything that may be understood by the light fantastic toe, the cup that cheers, the groaning board, the sound of revelry by night, and all that sort of thing according to the usual Christmas programme, but he had probably uppermost in his mind the fact that he had arranged to give an organ recital at one of the Cairo churches. One or two officers—Guy, for example, who was full out for animals of any kind—made straight for the Zoological Gardens, and spent hours there. Others, unable to throw off the fascinating bonds of their job, went to Heliopolis and renewed acquaintance with old friends in the other branch of the flying service. Others, again, tried to arrange for week-ends when there was racing on Saturday afternoons.

Perhaps the most curiously assorted pair were Photo-Bits and V.

Photo-Bits was never really happy unless he was doing something connected with photography. As to how he occupied himself before the war he was singularly reticent, and we could only guess that he must have spent quite a little of his spare time taking snap-shots. He had a number of specimens which appeared to show that he had been remarkably lucky in getting interesting places exactly at the moment when there were picturesque groups of figures posed about them. We looked with some curiosity at his camera to see whether it contained by chance some gadget or other which ours lacked, and we found it to be a heavy, clumsy sort of wooden box which he had designed himself. It contained ordinary dark slides instead of films, and allowed only for half a dozen plates. Any further supply had to be carried in a canvas satchel, and we grew hot at the thought of the porterage in a climate which made even the lightest of clothing irksome. I think that this weighed considerably with those photographers among us who might otherwise have struggled for the opportunity of accompanying him to Cairo in the hope of getting snap-shots as good as his. They didn't want to have to help to carry his apparatus.

V., who eventually joined him, guarded against this by supplying himself with a new camera of his own in addition to a small pocket one, so that if there should be any question about him having a spare hand, it could be definitely answered. He found, however, on arrival at Cairo, that P.-B. was not in any way dependent on casual assis-

THE BLUE MOSQUE, CAIRO.

tance. He engaged a guide to carry the camera and the reserve stock of dark slides, and here too his luck served him. By a pure accident—I am convinced there was nothing else in it—he got a guide named Mohammed Brown, who seemed

ON LEAVE

to know instinctively the kind of things P.-B. said he wanted, and he took him straight there. V. trailed along with the party, exposed plates when his companion did, and hoped for the best.

But it soon became evident to V. that P.-B.'s success with the camera was not all luck. In fact, very little of it was luck. The picturesque groups of figures which were the leading features of his collection were scarcely ever in position when they reached the spots to which Mohammed led them, and that was where Mohammed came in particularly useful. They would arrive, say, at a nicely carved doorway in the courtyard of an old Arab house. It would be inspected from all points of view. Then Mohammed, having been dispatched with various instructions into the neighbouring streets, would return leading a curious crowd of Arabs carrying chairs, tables and other properties. In a few minutes there would be a group playing chess or drinking coffee or doing something which just suited the back-ground. And then a photograph would be taken. And another. And still another. V. would stand stupefied at the reckless expenditure of plates until he began to understand that P.-B.'s "luck" consisted in selecting the best of perhaps half a dozen exposures made after an infinite deal of pains had been spent in arranging something that was worth an exposure.

At Cairo there was a very efficient military map-making department, and more than once "duty" took me there. I was thus able on several occasions to accompany P.-B. in his rambles. Once, piloted by Mohammed Brown, he obtained access to a little girls' school. It was situated in one of those delightful verandahs which, in the old streets of Cairo, appear at almost every corner. There was no difficulty in getting in; no difficulty in taking photographs, for P.-B. was not punctilious about holding up the education of young Egypt for an hour, and the schoolmaster was not averse from the idea of a "stand easy." A row of smiling little girls seated at a form with the teacher and a blackboard inscribed with Arabic writing in chalk formed the group which was eventually posed. The photographer was not concerned with the fact that the blackboard could only with the greatest difficulty be read by the students. That was a consideration far too pedantic. The touch of nature was what he wanted, and so overjoyed was he with his success, that he determined to take another view of the school from the outside with the little girls peeping out through the wooden railings of their schoolroom.

He went down into the street, and with a glance at the opposite houses, selected an upper window and knocked at a door. Mohammed Brown, as interpreter, was told that the window belonged to the owner's hareem, to which entrance was absolutely prohibited, unless—unless the officer would pay two shillings. P.-B., who was a Scotsman, figured this out as four times saxpence, and had another look at the window. He was not so sure that it would be really what he wanted, and Mohammed was commissioned to offer two saxpences. They were refused, and there was a period of inactivity during which Mohammed quickly disappeared, to return with a ladder. From the top of it, placed against the sill of the forbidden window, the photograph was taken, while the owner of the hareem watched with the expression of one who, grasping at the shadow, has lost the substance.

A story is told of Mr. T. P. O'Connor when he was starting his paper, *T. P.'s Weekly*. Instructing his staff as to the kind of thing he required, he drew their attention to a paragraph which had appeared relating to Zola who had recently been in England. Zola, it was said, in walking through the streets of London, was struck by nothing so much as by the number of hairpins dropped on the pavement. That, T. P. pointed out, indicated exactly the human touch he wanted, and in the office it became known as the " hairpin note." Photo-Bits with his camera had glimmerings of that sort of instinct. Cairo abounds in examples of exquisite architecture, but he cared very little for this unless he could procure a foreground of men, women, children, donkeys, camels, or goats, and though the majority of these accessories are very much a part of Cairo street life, they were seldom in evidence in the right quantities or qualities when

FOUNTAIN AND SCHOOL OF ABD ER-RAHMAN, CAIRO.

he arrived, on the scene. They had to be fetched, and having fetched them Mohammed Brown would enlist the services of a native policeman—a whisper and a furtive hand-clasp would do it—and the traffic was suspended until the show was over. Sometimes the properties and supers moved on in mass to another site, like a travelling circus, and the excitement continued for the best part of a morning; sometimes it was over in half an hour, and a handful of piastres rang down the curtain.

All this was a groping in the direction of the hairpin note. It did not always quite reach it, but the little girls' school got very near the real thing. It fired P.-B.'s enthusiasm, and hope ever held out the prospect of something as good or better. One day it seemed almost within his reach. He entered the University Mosque and found the shade of its great columned courtyard crowded with parties of little children, with their ink and reed pens and their polished tin tablets, seated round their instructors. There was a picture ready made, but here, alone I think of all the public places of Cairo, the camera was strictly prohibited. A visit to the dignified stately Arab principal of the institution was received with the utmost courtesy, but nothing less than a document signed by the responsible Minister of State could affect the restriction. A call was made at the Minister's office, and it seemed that the thing could be done. Next morning, at the hotel, an official appeared and undertook himself

IN THE SHARIA EL-GHURI, CAIRO.

ON LEAVE

to conduct the party to the University. But he had no written word and the embargo could not be removed.

Photo-Bits was in despair, but the official suggested that a picture as good could be taken in another mosque. He led the way to one of the most celebrated, and it was empty. Children, however, were to be had, and were brought. Also an Arab, as stately and dignified in bearing as the principal of the University himself, was persuaded to play the part of schoolmaster. A position was selected, but the light was unsatisfactory. Some fine specimens of the typical tent-work shaded it, and it was hinted that they might be drawn aside. To the official, a hint was an order. He procured a ladder, and before you could say "knife," he had borrowed one and was hacking at the ropes by which the big canvas squares were suspended. They came tumbling to the ground, the desired light was admitted, and the group was posed.

When the photograph had been taken, there came the question of remuneration. Half a piastre apiece was ample for the children, but for the grey-bearded and turbaned patriarch who had represented learning it was a more delicate matter. P.-B. tried him with five piastres; not P.-B. in person, of course, but P.-B. through his deputy, Mohammed Brown. The silver coin was rejected with scorn, and the Arab stalked away as though money were a thing of supreme indifference. It was noticeable, however, that he did not leave the scene of action altogether, but remained lurking in the shadows far away in a distant corner of the huge building. It was clear that, though withdrawn from the society of those whose ideas were evidently low and sordid, he was not so irrevocably withdrawn that an inviting voice talking of higher things—that is to say, in rather higher figures—might induce him condescendingly to return to it. The photographer felt in a generous mood. After all, the old gentleman had been the making of a promising picture. Why not do the thing handsomely? He sent Mohammed Brown into the shadows with an additional two piastres —one shilling and fivepence halfpenny in all —and the shadows seemed to light up with a ray of fervent gratitude. The gift had been accepted and there was peace.

Travelling about Cairo in this way, P.-B. saw most of what could be seen in the limited time at his disposal. But he was not the tourist —" rubberneck," as Guy would have said —that the rest of us were. P.-B. looked at Cairo only through the lens of his camera, and there are many things there in its broad thoroughfares and gloomy, narrow by-ways that could not be translated into the clear tones of a photographic print.

Had any one wished, for example, to photograph the staff of any of the principal hotels, he would have found his subject singularly elusive. Not that there was any lack of servants; quite the reverse.

ON LEAVE

But they were all specialists, and the particular specialist for your instant needs was never in evidence. There were, I think, four men at least whose duty it was to answer the bell of each bedroom. One made the beds, one attended to water, one cleaned boots, and one supplied refreshments. Not knowing this, the newcomer would ring the bell overnight and order tea to be brought to him at, say, seven in the

THE HOUSE OF THE JUDGE, CAIRO.

morning. He would wake at a quarter past eight, find no tea, and ring for it. He might get it in the course of half an hour. He was hot to know that he had given his order to the bed-man, and that it was not until he had expended the wealth of his vocabulary on the boot-man and the water-man that it had occurred to one or other of them to pass on to the refreshment man the glad news of his requirements. Thus the stay at an hotel afforded a pleasant change from the clockwork orderliness of the island routine, and only when one was on the brink of departure did it become clear how really lavishly one had been served. At that moment, in all the hurry and bustle of final arrangements, there would be a staff parade of self-conscious looking menials. Then, perhaps, a fully representative photograph might have been taken, but one would have felt little inclined to take it. Remembering how few were the hands that had offered to carry your bag from the cab to the entrance hall, it was marvellous how many there were to do the opposite. Considering the number of times you had grown tired of waiting for the lift and walked up the staircase, it was amazing how readily the lift attendant recognised you and how engagingly he smiled. Such things are, of course, noticeable in all hotels everywhere, but in Egypt I fancy they may be studied at their extreme. Throughout the land the keynote of all inter-racial negotiations is "backsheesh," and backsheesh means that while virtue may be its own reward, everything else has a price.

One gets as a visitor to Cairo the first hint of this in the Dragoman. Guides invite your patronage at the entrance of every hotel and on the kerb of every important street corner. One wonders how so many of them live. Among our officers there was one who had a disconcerting way with them if he did not require their attendance. He would not, as some would, reject their services with a contemptuous *imshi!* the word with which an Arab shoos away a dog. He would beckon the most insistent of them to approach and ask him in a friendly way if he wished, to go to the Pyramids. Nothing of course could please the man more. Our officer would then call up another from the group who stood watching in envy of their companion's good luck. He would make an elaborate ceremony of introducing the one to the other, and to the first he would say, " Ali will conduct you to the Pyramids. He knows all about them. Ali very good guide." It would take perhaps a minute for the significance of this to be fully appreciated, and a minute was ample for a retreat into a cab.

An alternative method was to accost one of the guides, say on the steps of Shepheard's Hotel, and ask him if he knew where Shepheard's Hotel was. The man with a smile, mingled with surprise and pity, would indicate that they were even then there. " If you will wait here then, for two hours, you may see me coming back. But I may be a little late." This equally with the other scheme gave time for escape.

Another officer—this was E. A.—despised guides of all kinds. There was hardly one of us who, on first arriving in Egypt, did not buy a little book called *Arabic Self-Taught*. E. A. was the only one, so far as I know, who ever taught himself more than the few phrases necessary for our dealings with the native working party on the island. But he did more than try to learn to speak Arabic. He devoted his spare time to learning to read and to write it, and he filled note-book after note-book with inscriptions in what is probably the most decorative looking language in the world, though, spoken, it sounds as though those using it are constantly out of breath. The attraction which drew him to Cairo was the hope that he would be able to find there old copies of the Koran in illuminated manuscript, and he had some little success, though nothing that was purchasable could bear comparison with the marvellous specimens in the Khedevial Library at the Arab Museum. His objection to guides was not so much that they were inevitably of no assistance to him in his antiquarian quest; it was not even that they showed him at every step that they were far prouder of their English than they were of his Arabic. It was that he simply didn't want them, so he would wave them aside and stand in the street and shout authoritatively and rather defiantly, *Arabaghi!* Some alert little vagabond would immediately reply, "Yes, Mr. Officer, kerritch," and summoned by the boy, the carriage would arrive. There was no need for a guide when you could do that sort of thing.

But the majority of us would have a guide when exploring the native quarter of the town, at any rate for the first time. Lacking E. A.'s linguistic

ON LEAVE

equipment one found it, in fact, rather difficult to avoid having one, because just at the start one was so obviously lost. It was a simple thing to shake them off at the hotel and drive to the Mouski—that long modernised thoroughfare so disappointing to those who have read accounts of it some years old—and one needed no guide to find the neighbourhood of the bazaars towards the far end of it. But in the bazaars themselves, walking was the only way, and no street map was much help. Alighting from the cab one stood for a moment bewildered, and it was here that guides, on the watch for such crises, presented themselves in vociferous and insistent abundance. On the whole it saved trouble to select one—you would be followed by three or four if you did not—and an hour or two of his company, brightened by his queer guide-book English, was a novelty quite worth the small sum that it cost.

An afternoon with a good guide would be enough to make one acquainted with the localities in which native life is to be found. Any one pressed for time, as the greater number of us were, might see a fair part of it by exploring the by-ways of the tortuous street crossing the Mouski and connecting the two old gateways, the Bâb el-Foutouh and the Bâb el-Zouweileh, and a little beyond for the bazaar of the tent-makers. In this street and its wandering tributaries, many of them reached by narrow doorways so like the entranc-

IN THE KHAN EL-KHALILI, CAIRO.

ON LEAVE

es to private premises that the unguided stranger would scarcely have the temerity to intrude, one came upon all manner of queer anachronisms. Eastern customs, costumes and industries which one might judge to have been but little affected by the progress of many centuries appeared side by side with the same things adjusted by the hand of the Western invader. And they seemed to flourish together, each in their degree, with few signs of that process of attrition which is called the survival of the fittest. Possibly each extreme really was the fittest for its particular purpose. East and West had both to be served, and there was in addition a curious blend of the two which formed a separating no less than a connecting link. Tumbledown native provision shops displayed their unattractive litter of wares face to face with twentieth century stores decked out with order and cleanliness. The Arab housewife might make her choice between a native-made clumsy tin pot or pan and a neat imported enamelled iron one. Native garments hung with garments of European cut, and articles from both wardrobes were frequently seen being worn by the same individual. Some of the workers in precious metals shut up their stock over-night in massive steel safes; others appeared to rely with implicit faith on the protection of the Prophet.

Much that was seen in a first visit to the bazaars left only a confused impression of extraordinarily confined passages with quite tiny square workshops on each side. They were most of them about as small as they could possibly be—some of them seemed smaller than they could possibly be—to permit of, perhaps one or two, perhaps half a dozen, men working. A certain amount of stock was stored there, and there was just room for a possible purchaser to sit on the threshold and drink coffee or weak tea while he bargained over some article which would be displayed in the pathway. One saw there workers in gold, silver and brass; shoe-makers; silk weavers; merchants of all kinds of spices, clothing and household goods; dealers in precious stones, antiques, carpets, trinkets. One passed through it all too quickly to remember details, but every now and then one emerged from the passages into spaces slightly more open and there were glimpses of old stone gateways, overhanging rafters, and the precipice-like sides of houses stretching up into the sky with their projecting lattice windows and their quaint irregularities. These more open spaces served the clearer-headed visitor as a kind of mental landmark punctuating his progress. And if he were not very clear-headed his progress might be punctuated, too, by gaps in the contents of his purse. The best guide, equally with the worst, would steer his convoy with the object of making him buy something, since it was, we heard, an understood thing that there was a 10 per cent commission for him if a sale were effected, and though guides are not highly educated, they

GATEWAY IN THE KHAN EL-KHALILI, CAIRO.

know enough arithmetic to realise that the bigger the price the bigger the commission.

Certain spots offered special opportunities for this class of traffic, and the stranger in his first tour must come upon them unexpectedly, one of these spots was a small shop in the Khan el-Khalili, about 10 feet by 8 feet, with a front open to the pathway. It was kept by a youngish and comely lady who sold laces and embroidered silks which were described as hand-work from native hareems. People who knew about such things did not, of course, impeach the lady's veracity, but they hinted that the hareems must have been situated in Europe or Japan or India, and that a machine had intervened between the hand and the fabric. I am no judge of such matters. I know that I passed that little shop many times, and the seating accommodation—just sufficient for two officers, not necessarily R.N.A.S.—was never empty, and I heard stories of a small fortune made during the war. But even in the Khan el-Khalili there may be such a thing as trade jealousy.

Another danger spot—if I may so describe them—was the shop of the perfume merchant who sold concentrated essence of the scent of roses, violets, jasmine, and lotus, and the exhilarating oriental drug extracted from ambergris. The pilotage of your guide and the blandishments of the proprietor would get you into the shop. There, in spite of your protests, the assistant would stroke your coat-sleeve or your handkerchief with the stoppers of his bottles. No more than this was necessary to convey a recognition of their contents vivid enough to convince you that never again would you be sufficiently defumigated for ordinary society. You could not prevent this intimate demonstration, and it was difficult to escape without buying at least a box of amber cigarettes. Indeed, it must have been nearly impossible, for cabins on the island were impregnated with amber for many days after leave was over. And it was an impregnation which did not restrict itself to the confined area of its origin. It came over the partitions and through them.

The first visit paid to the Khan el-Khalili under expert guidance would thus leave in the mind an impression recalling the story of the spider and the fly. One might hesitate to renew the experience, but if one did renew it—and it was well worth while—one found that the superficial glamour of the bazaars had worn off a little. One was no longer bewildered by the profusion of the wares of which the great majority at the initial rapid survey had seemed to be essentially Oriental and unexpected. Gradually, as one grew accustomed to it, the medley resolved itself into more definite and intelligible details, and the interest became centred at different points. It was possible to eliminate from consideration certain shops, chiefly Indian, which were merely of the catch-penny kind dressed out for tourists and identical in their display with a good many in Port Said. The labyrinth—as it seemed—of the bazaars became familiar, and one could walk with confidence through the intricacies of the quarters where was being carried on the real native work destined for the natives themselves, as distinct from the shoddy faked stuff—the spider's web—which was being made by way of advertisement and as a bait for the foreign passer-by. Notable shops became recognisable as places where there were good things always to be seen without any too embarrassing solicitations to purchase, and of these it became clear that some were outrageous in their prices and expected much chaffering and haggling, while others were reasonable and likely places for bargains.

There are in Cairo two chief industries which I think may be regarded as depending almost solely on tourists. There are minor ones, but these two are the most prominent. One of them is the engraving and chasing and decorating with silver and copper of all kinds of brass ornaments and utensils. The other is the making of the applique embroidery known as " tent-work." There is a third local industry, the inlaid work in wood, ivory and mother-of-pearl, but the tourist souvenir articles of this kind come chiefly from Damascus, and the Cairene work, which is applied largely to the ornamentation of modern tombs and mosques, is of a far finer quality of workmanship than any that

BRASS-WORK, CAIRO.

is expended as a rule on the cheap merchandise of the bazaars. One of the minor industries is in imitation of the wooden lattice work seen in the handsome projecting windows of old Arab houses, and with this one finds combined a certain amount of inlaid " mosaic." But here again Damascus and Syria are for the most part responsible.

In regard to the metal work, the R.N.A.S. explorers of the bazaars—and, I daresay, other officers from elsewhere as well—acquired considerable sagacity. It was evident, indeed, to any one who was concerned even with such metal working as that with which air-mechanics have to do, that the engraving and enrichment of brass with patterns in silver and copper which could be seen in process of manufacture in the Khan el-Khalili were largely of the clumsiest and most incompetent order. The Egyptian craftsmen had little or no notion of finish, and the designs by which they were guided were frequently poor in the extreme. A not too minute examination of the completed articles showed that this class of production resolved itself into two grades, and one found on inquiry that the superior grade had come from Damascus before the war, and that the inferior was that made on the spot. But every now and then one lighted with luck upon pieces which were infinitely superior to the recent importations from Damascus. These were secondhand, and had been made in Cairo or Damascus forty or fifty years ago.

ON LEAVE

The designs, consisting of interlacing geometrical patterns, foliage, and Arabic inscriptions—never of Egyptian figures such as were supposed to give the right touch of local colour to the cheaper varieties—were beautifully intricate and wonderfully carried out. There were generally in the best pieces a few little scrolls or letters in gold, mingled with the silver and copper. These were the pieces to buy, and more than one of them found its way to the Seaplane Base.

The embodiment in the designs of human figures and animals copied from ancient Egyptian mural paintings was a modern debasement of the pure Arabesque introduced, doubtless, to meet the tourist demand for something unmistakably Egyptian. The same thing was noticeable in the tent work which, if one be a stickler for archaeological accuracy, should have only such ornament as is to be found in mosques, where, indeed, there are models in endless variety. The quaint figure subjects in profile, the birds and the scarabs were generally preferred by purchasers from the Seaplane Base. After all, flying is a very modern thing, and flying men are not much given to archaeology. Here again one needed to choose with circumspection. There were little pieces of tent-work, and big; squares small enough (though a thought harsh to the touch) for a cushion, and pieces many feet long and wide, like those which were hacked down for the convenience of our photographic adventurer,

P.-B. You could pay high prices for the small pieces, and low for the large. It was a matter of care in the sewing and of the fastness of the colours. All kinds came to the island. We knew more about metal than we knew about needlework.

With the majority of our purchases, indeed, we were very much in the hands of the salesman. C. J., for instance, returned from leave with a hookah, which he knew to be intended for smoking. Confessing complete ignorance beyond this point, he had obtained, as he believed, the fullest instructions from the merchant together with a supply of tobacco and rose-water. Equipped with his acquired knowledge and the necessary implements, he appeared in the ante-room on the evening after his rejoining. Pictures which he had seen of hookahs in action all showed the operator seated on low cushions. The ante-room had no suitable low cushions, so he sat on the floor with the hookah, which he had brought in ready charged with tobacco and rose-water, standing at his side. There was a considerable audience from which, as with conjuring entertainments, voluntary assistance was readily available. C. J. issued his orders and a small live coal, the correct igniter of the special tobacco (which had all the appearance of having been prepared by crumbling a cigar), was applied to the small terra-cotta crater at the summit. Simultaneously he sucked at the big amber nozzle of the connecting hose. We heard the quiet purr of the bubbles in the rose water, but we looked in vain for smoke to issue from the performer's mouth. Fresh orders were issued. Puck was requested to remove his foot from the coils of the tubing, and new live coal was fished from the stove. This time there were three or four assistant stokers, and in their zeal they completely surrounded the stoke-hole. When the crowd at length drew aside, an ominous blue flame flickered high above the tobacco, and C. J. was seen to be emitting after strenuous efforts, small puffs of smoke. But perceiving the blue flame, he recognised that it was not part of the programme laid down by his mentor in Cairo. There was evidently something amiss, and there were various surmises to account for it—" The machine's back-firing"; "You've got a dud magneto," and so on. But C. J. was not satisfied until it was suggested that he had been blowing the wrong way, and had set fire to the rose water. He hailed this as the true solution, and took his apparatus to his cabin till such time as he could make all clear for a fresh start. While he was gone the stokers removed, lest the sight of them should convey a false impression, a tumbler of petrol and a fountain-pen filler.

Experts in architecture were, I think, rarer among us than experts in needlework. We could erect and furnish all sorts of fairly comfortable dwelling-places with all sorts of odd materials, even as our remote predecessors in exile made bricks

OLD CAIRO AND THE CITADEL.

without straw. The stove, for instance, which supplied the light for the hookah was made from an old oil drum. It was not much to look at, but it was highly efficient at its job. Yet though we lacked profound knowledge of the arts, we looked with a great deal of interest at much of what Cairo had to show by way of ancient monuments. We visited the great Pyramid, as in duty bound, and some of us climbed it inside and out, wondering the while how long it would have taken our Arab working party on the island to erect it, and calculating too the range of the wireless outfit whose aerial stretched from the mast at the top to the desert below. We did a round of mosques, and we went to the Citadel to see that of Mohammed Ali, that example of meretricious gaudiness, and bought pieces of yellow alabaster of the pedlar outside the door. We looked from the terrace at the superb panorama of domes and minarets, and the plains beyond it, at the tombs of the Caliphs, and the Mokattam hills with their quarries, and descending again to the city, we wandered at large in the streets with their amazing profusion of picturesque relics of an effete magnificence. They say there are towns in Spain where palaces may be rented at the price of cottages. I should think that must be so in Cairo. In the poorest quarters exquisitely carved doorways led to the most poverty-stricken habitations, and everywhere the seemingly ruthless hand of the jerry builder was laying waste great gaping tracks. Perhaps there was a scheme in it all, as there is in Paris where, however, fine buildings replace the meaner ones; but the work of demolition looked ruthless to the uninitiated. And there was no mistake about the alliance of squalor with splendid old architecture.

There is an Egyptian state department which has the care of ancient monuments, and its care is very well bestowed so far as it goes. Religious architecture seems, perhaps rightly, to get the preference, and one saw many laudable examples of good restoration. But the domestic architecture seemed to be fast disappearing, one could discover in the streets few good specimens of the genuine mushrabieh—the lattice work of those fine overhanging windows. Collectors of the antique are snapping them up, and there is apparently no one public-spirited enough to overbid the foreign buyer and so to preserve the character of the old streets. The carved wood is ripped out and cheap trellis of crossed battens in window frames curiously reminiscent of French provincial towns fills the gaps. All this seemed to be viewed with indifference.

I came upon one case of incredible vandalism. In the splendid fourteenth century mosque of Sultan Hassan there are some doorways decorated with marble of two different colours dovetailed together. This is a common practice in Moslem architecture, but here there is a distinctive detail. The dovetailing, if for the sake of simplicity I may so describe the interlacing of two elaborate patterns, appears not merely on the face, but on the return surface as well. Now every carpenter who has dovetailed the corner of a box knows that he cannot have his dovetail showing on each of two adjacent sides. On one appears the front of the dovetail; on the other the end. But the lintels in the mosque of Sultan Hassan show the dovetail not only on the front, but also underneath. They present, therefore, what seems to be an impossibility unless the marble has somewhere or other a hidden join; or unless the mason has performed the very skilful feat of cutting his intersecting pieces not square with their surface but on an angle. This, I am afraid, cannot be made very clear to any one who is unfamiliar with such technical problems. Let it suffice for me to say that the doorways do offer a very baffling puzzle. One day a tourist had a big bet on the question of whether the marble was solid or secretly joined. Actually there are no secret joins. The bet was decided by smashing a section of the marble, and thus villainously mutilated it may now be seen. Some one, I suppose, got some backsheesh out of the transaction. In what other civilised country could such a thing have happened?

The reason for such things gradually dawned upon us as we walked through the streets of old Cairo. We were in the capital city of a people which had lost its sense of nationality. Its houses

FOUNTAIN OF SHEIK MOTAHHAR, CAIRO.

and its mosques told the story of old times of magnificence followed by recent ones in which a succession of rulers, vassals of Turkey, regarded sovereignty solely as a means to the gratification of personal ends and ambitions, and cared nothing for their subjects so long as they could extort from them taxes which left a handsome margin above that which was due to the Sultan. A nation so governed knows nothing of patriotism nor of any rule of life but that of which the example is set by those in authority. Egypt had for centuries been ready to receive the impress of any external influence strong enough to make an impression, and everything had its price.

I think it is in the architecture that this is to be most easily traced. The story is not very apparent to the superficial glance of the casual visitor who finds efficient state railways and flourishing hotels to serve him. Nor is it at first discernible in the glamour of the bazaars. But in the streets one may read it. There is no longer, as there was in the days of the Caliphs, a national style of design. In the new quarters the florid modern style of western Europe is everywhere. It is attractive, but it is not of the country. Where in the older parts old houses have been pulled down, they have been replaced by flat-fronted houses with no distinctive feature, excepting the narrow windows which remind one of French provincial towns. They are unassuming and perhaps not unsightly, but they

ABOVE AND BEYOND PALESTINE

are not Moslem, though those who dwell in them are. Everywhere the foreigner has had a free run. Nothing has stayed his hand.

It is not merely a matter of replacing old with new. That is happening in all towns with a history. But in countries where there is pride of race the new has a national character no less than the old. In Cairo, as it seemed, the West is being mingled with the East.

That the process will be ultimately beneficial is evident, for there are signs that after a long period of exploitation the Eastern element under Western encouragement is awakening to a sense of its individuality. This, too, may be read in the architecture. The railway station is an example of an interesting if not very successful attempt to adapt the Mohammedan style to a purpose not really suitable, and there are other and better instances. One hopes that the full awakening will come in time to save many memorials of the old time which now seem to be doomed. Meanwhile, Egypt is like a sick man recovering from a disease which has been intentionally prolonged and aggravated by the physicians for their own profit. The treatment is now changed, but the cure is yet very far from complete. Cairo shows many symptoms of the old regime and of the new. There was every reason for us to return from leave there to the Base at Port Said with wider views of the part which we had to bear in the work of salvation.

XII

THE METAMORPHOSIS AND THE END

It was the story of the Royal Naval Air Service in the eastern Mediterranean that I set out to tell, and of that, though nearly another year elapsed before the war drew to its close, there is little more to relate.

During 1918 the seaplanes were incessantly occupied with submarine patrols, not only from Port Said, but also from Alexandria, where another base had been formed early in 1918. These patrols were a very valuable but rather dull preventive measure, which produced on only four occasions the excitement of an actual sight of the enemy. The work was continuous, but as everybody knows, on 1 April, 1918, the R.N.A.S. combined with the R.F.C. and emerged as the R.A.F. Before this took place, there were two carrier expeditions, of which one took us to Mudros, a long way out of our own area of operations. The other was in our old hunting ground—the Red Sea.

It was about 20 January, 1918, that the news spread in Port Said that the Goeben and the Breslau had made a sortie from the Dardanelles.

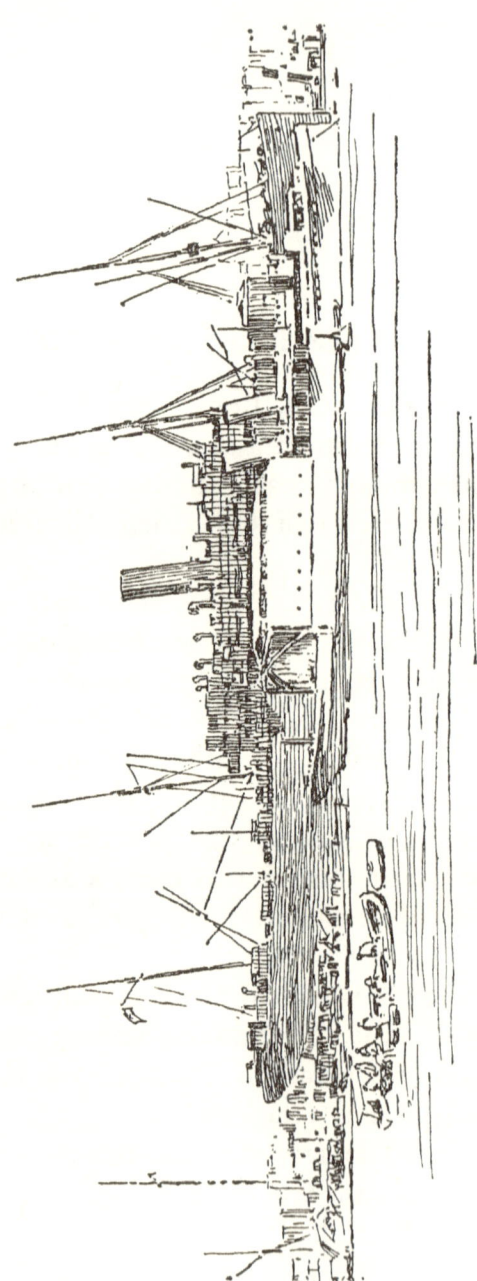

H.M.S. EMPRESS AT PORT SAID LYING ALONGSIDE THE MINNETONKA, AFTERWARDS SUNK BY SUBMARINE.

THE METAMORPHOSIS AND THE END

The Dardanelles are a far cry from Port Said, but the news sent a tremor of excitement through the town because the enemy was evidently bent on more mischief than could be done in the Greek Islands or at Salonika, and it seemed reasonable to suppose that the big guns of the Goeben might get to work on the two Egyptian ports. The intention, if it existed, came to nothing, for the Breslau was sunk before she had got any great distance from her own waters and the Goeben turned tail and tried to make her way back to Constantinople, only to run aground on Nagara Point in the Narrows.

It was then that urgent orders came for the Empress to proceed to Mudros to lend a hand with the bombing of the stranded armoured cruiser. She arrived on 24 January, sent out two seaplanes the same night, and obtained a direct hit, though both pilot and observer were doing a long flight in darkness for the first time. For two days the weather prevented further operations, but on the 27th another night flight was attempted. The machine did not return, and reports received later showed that both its occupants had been captured unhurt. Their machine had been brought down from a low altitude by machine-gun fire. Meanwhile, the Goeben had been floated off, her 3-inch plates little the worse, I am afraid, for our direct hits, and was out of reach of further attacks. The Empress got back to her base on 4 February.

The Red Sea trip provides little of special interest

for our story. Battleships were operating in the neighbourhood of Loheiya, and many flights were carried out with the objects which previous trips in this area had made familiar—spotting, bombing and reconnaissance, and demonstrations to impress the natives. The expedition started on 13 February, and the City of Oxford was back again on 29 March.

This was three days before the transformation. For months the constitution of the coming Royal Air Force had been discussed, and already there were experts among us who were prepared to explain the precise meaning of obscure passages in the printed memoranda which reached us. We all went ashore and celebrated the wake of the R.N.A.S. at a dinner at one of the hotels, and there was a ceremony on the following evening not distinguishable, from the other, to commemorate the birth of the R.A.F. In the interval between the two banquets we had ceased to fly the white ensign, and our personnel had gone to bed squadron commanders and flight lieutenants, and arisen majors and captains.

For the moment that was all the real difference it made to us, but gradually other changes became apparent. Generals began to share the interest in us which hitherto had largely belonged to admirals; our first lieutenant became an adjutant, and various problems of discipline and ceremonial presented themselves to the "ratings" who had now

THE METAMORPHOSIS AND THE END

become "other ranks." Chief of these, perhaps, was the question whether they should remove their caps before or after saluting. All this sort of thing, however, righted itself, and as the R.A.F. we settled down to our submarine patrols.

The spring and the summer of 1918 dragged on. Expert prophets gave the war at least another year, and we watched with keen though not very excited interest the swaying line on our map of the Western Front. There were rumours of approach-

MEJDEL YABA.

ing dissolution in our own quarter of the globe, but we took little heed of them. The Turks were said to be utterly exhausted, but they showed no signs of it in Syria. The same had been said of them before, and we thought little of it.

Then came General Allenby's amazing advance, a thing to everybody, excepting those in the innermost know, so utterly unexpected that it was almost incredible. With a surprise attack not dissimilar from that which resulted in the capture of Gaza, he had turned and broken the front which stretched from Mejdel Yaba, a little town near the coast between Jaffa and Haifa, to the hilly country north of Jerusalem and Jericho. We heard stories of whole Turkish divisions cut off and captured, of the advance on the right of our old friends the Arabs from the Hejaz, and of the complete rout of three Turkish armies. Names of towns which had long been familiar to us in connection with far distant seaplane operations appeared in the telegrams as having fallen into the hands of the advancing allied troops—Tul Keram and Nablus, Haifa and El Fule and Nazareth.

Where was it to stop? Would Beirut be taken, and Damascus? We had hardly time to wonder before their names were added to the list. It was stupendous, but it seemed to us almost like poaching! The advance was, indeed, so rapid, that the victors themselves nearly forgot that it was not really absolutely new country. An intelligence summary passed to us "for information" said that the first British aeroplane had just flown over Damascus. A manuscript note was added in the margin reminding the author that a seaplane had performed this feat on 28 February, 1917.

Meanwhile, Bulgaria had capitulated, Germany began negotiations for an armistice, Austria and Turkey followed Bulgaria, and the end came, when our daily seaplane convoy escorts were cancelled and Port Said lighthouse once more, after four years of blindness, flashed out its revolving rays.

And then we faced a period of utter stagnation.

THE METAMORPHOSIS AND THE END

There were the usual things to do ashore after a sort of make-believe round of duties had been carried out at the Base, but no one wanted to do them when they had ceased to be a respite from actual duties. Everywhere there was a rush to get demobilisation orders through in time for Christmas. Few, if any, of our community were successful.

To me in this dull period came one bright interval. On a certain day towards the end of November there appeared in the mess an officer of the Naval Reserve who explained that he was an artist with orders from the Admiralty to make pictorial records of the work of the Navy in those parts. Seaplanes fell within his terms of reference, and to me, as intelligence officer, he came for information. I loaded him with it, showed him photographs and let him glance at filed reports. It was easy to convince him that there was far more than he had believed to be in existence, more than he could get clearly into his head in the brief time at his disposal, for he had to hurry on to Mesopotamia. But it was the inspiration of the major commanding our squadron at Port Said that the only sound way was for me to accompany him to the actual spots. In an hour or two I had been provided with orders to proceed with the artist to an unspecified destination, and together we set out on a journey which lasted three weeks, and took us through much of the country over which our seaplanes had flown. Gaza, Askalon, Ramleh, Ludd, Tul Keram, El Fule,

ABOVE AND BEYOND PALESTINE

Haifa, Tyre, Sidon, Beirut and Damascus were all included legitimately in this amazing duty joy-ride, and we even dragged in Jerusalem, for which there was no excuse.

But the story of this journey has no place here.

PORT SAID.

CECIL ELDRED HUGHES
(1875–1941)
Biographical Afterword

Cecil Eldred Hughes was born on 3 December 1875, in Marylebone, London, the second of three boys, to Eldred Augustus Hughes, an architect and surveyor, and Jessie Maud Hughes, née Cobbett. His elder brother Augustus Edward had been born a year previously, and Ernest Theodore Cobbett in 1878. Augustus became an architect, Fellow of Royal Institute of British Architects, Freeman of London (1911) and Mayor of Marylebone (1922–23). He died, aged 100, in 1975. Whilst Ernest (d.1961) became a doctor and Member of the Royal College of Surgeons. In comparison, Cecil claimed to be just an author and journalist.

Cecil attended the Misses Thomson's Preparatory School, in Hove, Brighton, possibly at the same time as a more famous pupil, Winston Churchill, who remembered his time there fondly. 'At this school I was allowed to learn things which interested me: French, History, lots of Poetry by heart, and above all riding and swimming.' The school apparently had a wide-ranging curriculum as Cecil learnt drawing, piano, gymnastics and drill, also some Greek.[1] From Brighton, Cecil moved along the coast to Eastbourne to continue his education at Eastbourne College, 1890–1895. Here he joined his brother Augustus (1890–1892), and they welcomed Ernest a year later (1891–1896). Eastbourne College was an Independent School, that is, fee paying and free to set its own curriculum. It can be assumed that the teaching was strong on grammar, languages (ancient and modern), history (ancient and British), religion, and sports, but light on maths and sciences.

Leaving school Cecil attended University College London from June 1896. Sadly, no details of what he read or even whether he gained a degree have survived. However, and perhaps more important, he met and made a lifelong friend of William Wedgwood Benn. Benn's father, John Williams Benn (he was knighted in 1906), had been a London councillor and MP since 1889. His son intended to follow in his father's footsteps. Benn was president of the university's debating society, graduating in 1898 with first class honours in French. He was first elected to Parliament in 1906.

On leaving University College London, Cecil became private secretary to Sidney Lee (1859–1926, knighted 1911), and editor of the *Dictionary of National Biography* (1890–1917).[2] Sidney Lee was a Shakespeare scholar and encouraged Cecil

BIOGRAPHICAL AFTERWORD

to compile *The Praise of Shakespeare: An English Anthology*, to which he contributed a Preface. The book, published by Methuen in 1904, appears to be Cecil's first authorial work, although it was preceded in print by a little magazine article.

In April 1904 the first edition of *C.B. Fry's Magazine*, later *Fry's The Outdoor Magazine*, was published, to it Cecil contributed *Everyday Things We Do Wrong: Dorette Wilke*.[3] Shortly afterwards he became Fry's Assistant Editor in which post he remained until 1910 or 1911, also providing many articles, and poetry usually illustrated by his sketches.

Cecil's friendship with Wedgwood Benn remained close. Evidently spending time *en famille*, as, on 3 June 1905, Cecil married Lilian Margaret Benn. The wedding, apparently a 'very pretty one', took place at Blackheath Congregational Church in Lewisham, London. There were six bridesmaids but no mention of a best man. The couple honeymooned in the Austrian Tyrol. They had two children, a son Keon Eldred (b. 1906) and a daughter Sylvia Margaret (b. 1909).

During the ten years from 1904 Cecil wrote two more books to be published by Methuen. *A Book of the Black Forest* (1910) was illustrated with his sketches. His developing interest and expertise resulted in *Early English Water-Colour* (1913), also published in an American edition (A C McClurg & Co., 1914) and reprinted in 1929 by Methuen and 1950 by Ernest Benn. In addition to *Fry's Magazine*, he contributed to many other publications, including *The Globe and Traveller* newspaper, *The Harmsworth Red Magazine* and *Cassell's Family Magazine*. He also wrote book reviews in rhyme for *Punch*. They were mostly light-hearted, even playful as this example demonstrates:

> *My Merry Rockhurst*, latest born
>   Of E. and AGNES CASTLE's books,
> Tells of the Restoration's morn —
>   Go to! Oddsbodikins! Gadzooks!
>
> Messrs. SMITH, ELDER publish it;
>   KING CHARLES THE SECOND wanders through,
> Lax, dignified, a rake, a wit —
>   Oddsbodikins! Gadzooks! Go to!
>
> Thrills upon thrills in mad career
>   Keep moving, till the best man wins,
> All in the proper atmosphere —
>   Gadzooks! Go to! Oddsbodikins![4]

Whilst a review from 1906 may be seen as almost prescient in view of an absorbing interest of his post war years.

> To own a Wisley or a Kew
> May be too much for me or you;
> But everyone can dig and hoe
> And rake and weed and prune and sow
> (Especially on Saturday)

BIOGRAPHICAL AFTERWORD

A little plot, and acre say.
Now every small *jardinière*
Should straight to Mr. Curtis fare
For his *Small Garden Beautiful*,
A volume indispensable
(At Smith and Elder's, seven-and-six),
To set more peas a-climbing sticks,
To fill more beds with mignonette,
To make sweet England sweeter yet.[5]

By early 1909 Charles Fry was increasingly absent from the helm of his magazine, preferring to spend his time aboard an old barque converted to boys training ship *Mercury*, of which since 1908 he was Captain-Superintendent. In consequence Cecil became acting editor. He lacked the contacts of Fry and the magazine became more reliant on articles culled from other magazines. However, he struggled on but the publisher, Newnes, eventually had had enough and gave Fry an ultimatum — return to the editor's chair or resign. Fry chose to remain aboard ship, and Newnes ceased publishing the magazine in March 1911. A *New Fry's Magazine* was on the bookstalls a few weeks later, but by then Cecil had found new employment.

Once again the Benn connection came to the fore and Cecil became an employee of Benn Brothers, from 1923 Ernest Benn, Publishers, for the next thirty years, interrupted only by military service during the Great War. At this time, they were mainly publishers of trade journals and 'technical books for each specialized public.' Cecil was probably using his editing experience and skills in preparation of the trade magazines.

On 23 November 1914, Cecil applied for a Temporary Commission in the 9th Battalion, Somerset Light Infantry. He was one month shy of 39 years of age. On 9 December he was informed that his request had been granted. It is not clear why he selected the Somerset Light Infantry, there is nothing to indicate any previous connection with the county. However, he spent the next two years a Temporary Lieutenant, 'in training camps and depots doing home service jobs, of which some were not without interest.'

In 1916 he was suffering from an abscess in his right antrum, an infection of the sinus. For several months a Medical Board classified him 'Unfit GS, fit for Home Service.' A memo dated 24 November 1916 informed the Medical Board that, 'This officer is required as military observer Seaplane base Suez. He is serving at home & states he was passed fit for Home Service only & has now received orders to attend an MB at Birmingham & also orders to proceed to Egypt tomorrow.' A Medical Board on the same date found that whilst Cecil was still suffering from the abscess, 'he keeps it clean by syringing out the cavity. He is fit for duty as military observer Seaplane base Suez.' Lieutenant Cecil Eldred Hughes, Somerset Light

BIOGRAPHICAL AFTERWORD

Infantry, arrived in Port Said shortly before Christmas 1916. His experiences are described in this book.

LIEUT. C.E. HUGHES IN 1918 WHEN SERVING WITH THE EAST INDIES AND EGYPT SEAPLANE SQUADRON.

But why was this relatively old Lieutenant, with no previous aviation experience, suddenly in demand as a military observer with a Royal Navy seaplane squadron? There was no shortage of willing volunteers already in country. It may be not insignificant that a certain Captain William Wedgwood Benn had been serving with the squadron as chief observer and intelligence officer, for some months. It seems likely that an exchange of letters created the desire and some string pulling by Benn in Egypt created the opening.

Cecil remained in Port Said until 'Transferred to the Unemployed List' on 24 February 1919.

A SHORT FLOATPLANE
APPROACHING HMS EMPRESS OFF BEIRUT
SEPTEMBER 1917.

BIOGRAPHICAL AFTERWORD

Initially, he was Assistant Intelligence Officer, under Benn, to the East Indies and Egypt Seaplane Squadron, RNAS. He took over the job in May when Benn returned to England. The job gave him a unique view behind the scenes as well as from the front line, although there is no record of him flying operationally as an observer. On 1 April 1918, he was transferred to the RAF, continuing as Intelligence Officer with 269 Squadron, RAF. For his services he was twice Mentioned in Despatches, 2 July 1918 and 3 June 1919. His unemployment took effect on 1 April 1919 upon his return to England. He was not unemployed for long, returning immediately to his job at Benn Brothers.

In 1921, on William Wedgwood Benn's recommendation, Ernest Benn hired Victor Gollancz, who expanded the company's publishing into art books and novels, especially children's books. Cecil was chiefly responsible for the production of a series of 'the most expensively produced art books... magnificent books for the connoisseur.' Although Gollancz massively increased Benn's turnover, he was an uncomfortable fit in the politically conservative company. Gollancz was left leaning in politics and a supporter of socialist movements. He left in 1927 to form his own, successful company which survives to this day, mostly known for its post-WW2 move into Science Fiction publishing. It is possibly about this time that Cecil became a director of Ernest Benn publishing.

ABOVE AND BEYOND PALESTINE

For a short period after the war, Cecil and his family appear to have been guests in the Benn household. However, by 1920 they had purchased The Priory, Orpington, Kent, as a family home, together with 15 acres of adjacent wooded meadow land. The Priory has a 13th Century Hall with 'modern' additions—it survives to this day. Cecil had found the 'little plot, and acre' he referenced in 1906. He and his wife devoted considerable time to restoring and enlarging the gardens originally laid out in 1865, 'which they transformed from an open meadow into a thing of real beauty, with soft lawns and grass paths about formal flower beds and shrubberies. With these and the lofty trees for a perfect background, [they] presented outdoor plays on summer nights.' The Theatre Garden was the work of Geoffrey Jellicoe (ca.1927),[6] a family friend who designed the area with a raised grass platform and stone steps as seats for the audience. The theatre was used by the Hughes' family for outdoor entertainment. The Hughes' were an affluent family. In 1921 they employed five live-in servants—two gardeners, a cook, parlour maid, and house maid. In 1939 they employed a butler, cook, two maids, and a garden boy.

Back at Ernest Benn, Cecil continued his role as editor and director. He also found time to write. In 1924, for John Lane, he translated *Richard Parkes Bonington: His Life and Work* by A. Dubuisson, contributing extensive notes from his own researches on this watercolourist. In the same year he contributed a chapter on 'Water-Colours, Drawings and Prints' to *The Book of the Queen's [Mary] Dolls' House* (Methuen, 1924), it was also a recognition of his growing expertise on the subject of English water colour artists. Then in 1928 he wrote a slim paperback book for Benn's Sixpenny Library, on the subject of *The English Water Colour Painters*.

Cecil's final published book was *Above and Beyond Palestine*. The story of how it came to be published should be told. As he says in the Preface, the book was written in the final months of the war. It is possible that he brought home the completed manuscript which, having had the hoped-for cathartic effect, was put out of mind. In 1929, author Edward John Thompson, well-known for his fiction and non-fiction books on India, and an officer in the Leicestershire Regiment during the First World War, submitted a Palestine memoir to Ernest Benn. When Cecil showed Thompson his sketches from Palestine, it was quickly decided that they should illustrate Thompson's book. Cecil mentioned *inter alia* his manuscript. After reading it, Thompson recommended that Cecil's memoir be published together with more sketches.

When published in 1930, *Above and Beyond Palestine* appears to have had a lukewarm reception. Perhaps the reviewers, and public, were tired of the endless war stories. However, one reviewer in *The Spectator* caught the spirit in which it was written.

ABOVE AND BEYOND PALESTINE

'The work of the naval airmen in the Eastern Mediterranean and the Red Sea during the War has never been described until now. How incessant and effective it was we may see in Mr. C. E. Hughes' clever and amusing book, *Above and Beyond Palestine* (Benn, 10s. 6d.), which he has illustrated with many attractive sketches in line. Long before our troops entered Palestine, our seaplanes had been mapping the country and worrying the Turk by raids on his communications and depots, to say nothing of combined naval and air attacks on Beirut and other places. Mr. Hughes is not concerned to write a history of this side-show, but rather to give some idea of how the East Indies and Egypt Seaplane Squadron was employed, how its officers lived on their island at Port Said, and how they amused themselves on leave in Cairo. Few books about the War have been so cheerful or so brightly written as this modest volume.'

Cecil continued as a director of Benn's and tending his gardens until his death in 1941. He suffered a heart attack in December 1940 whilst filling a bomb crater in the Priory's drive way. He never fully recovered and passed away on 19 July 1941.

After Cecil died his widow continued to live at The Priory. In 1947 she sold the building to Orpington District Council for use as offices. The Council purchased the grounds in 1959. After an extensive programme of improvements, the grounds were opened as a public park in 1962.

BIOGRAPHICAL AFTERWORD

In 1959, the south wing of The Priory was demolished and a new public library constructed on the site. The Orpington Museum opened in The Priory in 1965, moving to a new site in 2015. In 2016, V22, a company specialising in offering artists' studio space and contemporary art exhibitions, was awarded a 125-year lease to manage Orpington Priory. The Priory Gardens and Park remain a public space, the gardens containing a dedication to the memory of Cecil and Lilian Hughes.

IAN M. BURNS

TORONTO, *February*, 2025.

ENDNOTES

1 Eastbourne College, Admission Form, 1890.
2 A C.E.H., aka Charles Ernest Hughes (1867–1933), English clergyman, contributed to *Dictionary of National Biography* (DNB) 1888–1900, this has incorrectly been taken to mean Cecil Eldred Hughes. There is no evidence that Cecil contributed directly to the DNB.
3 Dorette Wilke, founder of the Chelsea College of Physical Education which trained women gymnastic teachers.
4 *Punch*, 27 November 1907. Agnes and Egerton Castle, *My Merry Rockhurst*, 'being some episodes in the life of Viscount Rockhurst, a friend of King Charles II, and at one time constable of His Majesty's tower of London.'
5 *Punch*, 30 May 1906. A.C. Curtis, *The Small Garden Beautiful*, 'a beautifully illustrated and practical guide for creating and maintaining a stunning small garden.'
6 Later knighted, Sir Geoffrey was not related to Admiral of the Fleet John Rushworth Jellicoe, 1st Earl Jellicoe.

FLOATPLANES OVER THE DESERT

by Ian M. Burns

---

*The Adventures of
French and British Naval Airmen
Over Sea & Desert Sand
1914–1918*

AVAILABLE FROM LITTLE GULLY
IN 2025

littlegully.com

www.ingramcontent.com/pod-product-compliance
Lightning Source LLC
Chambersburg PA
CBHW060552080526
44585CB00013B/533